CHARGING AHEAD!

BY THE SAME AUTHORS

The Bonjour Effect, New York, St. Martin's Press, 2016.

The Story of Spanish, New York, St. Martin's Press, 2013.

The Story of French, Knopf Canada, St. Martin's Press et Robson, 2006.

Sixty Million Frenchmen Can't Be Wrong, Naperville (Illinois), Sourcebooks, 2003.

Jean-Benoît Nadeau and Julie Barlow

CHARGING AHEAD!
Hydro-Québec and the Future of Electricity

Translated by Julie Barlow

Baraka
Books
Montréal

ISBN 978-1-77186-201-1 pbk; 978-1-77186-209-7 epub; 978-1-77186-210-3 pdf

Cover: Designed from images of Ackkasit\shutterstock.com
Book Design by Folio infographie (with Claudia McArthur and Nathalie Caron)
Editing and proofreading by Aleshia Jensen and Robin Philpot

Translated from the French by Julie Barlow

Original title: *Branchée. Hydro-Québec et le futur de l'électricité*
© 2019, Éditions Québec-Amérique inc. All rights reserved.

Legal Deposit, 4th quarter 2019

Bibliothèque et Archives nationales du Québec
Library and Archives Canada

Published by Baraka Books of Montreal
6977, rue Lacroix
Montréal, Québec H4E 2V4
Telephone: 514 808-8504
info@barakabooks.com

Printed and bound in Quebec
Trade Distribution & Returns
Canada and the United States
Independent Publishers Group
1-800-888-4741 (IPG1);
orders@ipgbook.com

We acknowledge the support from the Société de développement des entreprises culturelles (SODEC) and the Government of Quebec tax credit for book publishing administered by SODEC.

The authors would like to thank Hydro-Québec for its support during the research of this book. The content of the book is entirely the authors' point of view and opinion.

Contents

Major facilities

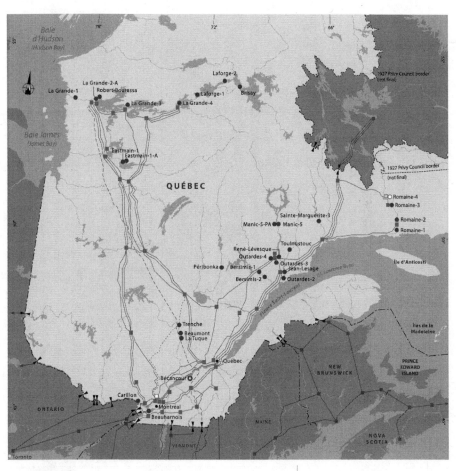

Generating station rated 245 MW or more	●	Hydro
	◐	Thermal
Other facilities	○	Generating station under construction
	■	735-kV substation
	□	735-kV substation under construction
	——	735-kV line
	·····	735-kV line under construction
	– – –	450-kV direct-current line
	►◄	Interconnection
	—■—	Neighboring system (simplified)

Introduction

The Future of Electricity

Imagine a future when Quebec's four million house-holds will all be equipped with interconnected solar panels and batteries. Home owners will turn off lights, open curtains, lower the temperature of their homes and manage all their energy needs with a smartphone. The combined energy these houses produce and store will create a gigantic, decentralized stockpile of electricity.

But that's not all. Outside, on Quebec's roads and highways—but also in China and Europe—millions of buses and delivery trucks will be fitted with Hydro-Québec motors that run on made-in-Quebec batteries. In this future, all the thermal power plants between Boston and Toronto that now run on coal or oil will be shuttered and instead, city centres will be lit by a "Super-Hydro-Québec." Back at Hydro-Québec's headquarters in Montreal, ambitious employees will be waiting for job transfers to one of the company's new offices in London, Mexico City, Paris or even Beijing.

Hydro-Québec has everything it needs to turn this into a reality in the next 25 years. On the 75th anniversary of its creation, Hydro-Québec has become the second-biggest hydroelectricity system in the world after China Yangtze Power. It is the only power system in North America with its own research centre, one that puts it on the cutting edge of technological developments in the field of electricity. Thanks to Hydro-Québec's energy sources, which are considered 99.8 percent renewable, few power systems in the world outside of Norway and Iceland can claim to be "greener." And no other system on the continent has lower rates. Hydro-Québec's reputation attracts delegations from China, Mexico, the U.S., Africa and Europe who flock to its generating stations, control centres and laboratories to discover new research and technology coming from Quebec.

Yet there is another, darker vision of Hydro-Québec's future that could also become a reality. Hydro-Québec's President and CEO Éric Martel alluded to this in January 2018 when he told the *Journal de Québec* that the company could face the threat of a "death spiral."[1] It was probably the first time most Quebeckers ever heard the expression, or stopped to think that the crown jewel of their economy might be in danger.

Death spiral conjures up a strong image of a problem that's actually threatening most of the power systems on the continent. Power rates across North America have risen so much that consumers are starting to look for ways to reduce their energy costs, like installing solar panels in their homes and becoming self-generators. But power systems are paying the price for this: as customers turn to self-generation and demand falls, utilities sell less electricity and make less money. But their costs haven't dropped. So to compensate, they either raise their rates or ask the government to pick up the tab (with taxpayer dollars). The cycle then becomes self-perpetuating. Rate increases push more customers toward self-generation, which shrinks utilities' revenues further and forces them, once again, to increase their rates. The syndrome is already affecting power systems in Ontario, California and Hawaii. Caught in a vicious cycle of declining demand, systems that fail to react quickly or make the wrong choices, run the risk of suffering serious damage or disappearing entirely.

The problem is still a bit hypothetical in Quebec where, in 2018, there were still only a total of 716 households producing their own energy. But the rate of self-generation in Quebec is only low because Quebeckers benefit from the lowest hydroelectric rates on the continent and at the moment, solar panels, windmills and batteries are still more expensive energy alternatives. This will change. Hydro-Québec itself predicts that in 6 or 7 years, perhaps as early as 2023, photovoltaic technology will have vastly improved, and the price of solar panels and windmills will fall to an affordable level. The shift may have already started. In 2018, after years of incremental increases, the number of energy self-producers in Quebec suddenly almost *quintupled*. If the trend of self-producing takes off in Quebec, Hydro-Québec, like power systems elsewhere on the continent, will face the possibility of losing its monopoly—a first in the history of the state-owned company.

As professional magazine journalists, we have interviewed and written about Hydro-Québec several dozen times over the last two decades for stories on Hydro-Québec and on hydroelectricity and energy in general. We got the idea for writing a book on Hydro-Québec in 2016 when we were each working on separate energy-related stories. Jean-Benoît was writing a profile of Éric Martel, who had been named Hydro-Québec's President and CEO the previous year. During the interview, Martel explained his plans for Hydro-Québec to expand into the international energy market and told Jean-Benoît how he hoped to assuage the growing defiance Quebeckers were showing toward their energy utility. Julie, meanwhile, was writing an article on exactly that topic. During research for a story on challenges to preserve Quebec's landscapes Julie interviewed Lisette Lapointe, who was then the mayor of the town of Saint-Adolphe-d'Howard in Quebec's Laurentians region. The town was in the throes of a battle trying to force Hydro-Québec to stop building a high-voltage power line that would cross its territory. While writing the articles, we realized they highlighted some of the biggest challenges Hydro-Québec was facing. Jean-Benoît discovered that because demand for electricity in Quebec had stagnated, Hydro-Québec had to find new places and new ways to sell its electricity—in short, it needed new markets. Meanwhile, Julie was exploring how changing consumer mentalities were making it harder than ever for Hydro-Québec to get local populations to let them build new lines—which Hydro-Québec would obviously need to expand into export markets. In other words, Quebec's population was going to make it hard for Hydro-Québec to expand, or maybe even survive in the long run.

Our first idea was to write a history of Hydro-Québec commemorating the 75th anniversary of the nationalization of electricity. However, when we started the research, we quickly realized that Hydro-Québec's future was a more interesting topic. Since Thomas Edison first demonstrated his light bulb in 1879, electricity has been generated in a familiar way: methods have just gotten better, more efficient and yielded more energy. But that was only true until the last decade. Since then, everything has changed: not only technology and energy demands, but also the way people even think about energy. These changes have thrown into question decisions made by all power systems on the planet—even the best managed among them—and shown that many past assumptions about how the future of energy would unfold were simply wrong. In short, as Hydro-Québec was approaching its 75th anniversary, we realized that while in many ways it is the same company it was on its 50th anniversary, by its 100th anniversary, in 2044, it will be very different. Hydro-Québec is entering an era of change that will not only transform it, but also reshape the future of electricity everywhere it's produced.

The Hydro-Québec of 2044 won't have much to do with the industrious beaver featured on the company's original logo, created after the war (Fig O-A). Hydro-Québec will never again build new dams at the pace it did in the 20th century. Hydroelectric dams may be simply replaced by millions of solar panels, windmills and batteries, for that matter. These new tools, coupled with advanced home automation technology and energy effi-

Fig. 0-A: Hydro-Québec's logo from 1944. The beaver and the Maple Leaf were French Canadian symbols at the time. The Fleur-de-Lys and British lion were partly hidden behind the lettering.

ciency measures, could constitute a new kind of energy supply comparable in capacity to the huge reservoirs of Hydro-Québec's power dams today. Hydro-Québec could even manage these new reservoirs remotely, like it already does with the water supply in its dams. Together, these developments will have the effect of lowering hydroelectric requirements in Quebec, which, in turn, will free up electricity that can be sold in Toronto, New York, Boston or Halifax.

In the summer of 2018, when we were in the middle of writing this book, we packed up our daughters and took them on a road trip along the North Shore of the St. Lawrence River to see Manic 5. Although it is officially called the Daniel-Johnson Dam, all Quebeckers know it by its original name. This spectacular dam located 220 km north of Baie-Comeau is the emblem of Quebec's vast hydroelectric project. Built between 1959 and 2005, the nine power stations that capture the power of the rivers Aux Outardes and Manicouagan, together, constitute the second-largest hydroelectric complex in Quebec after James Bay (which is officially the La Grande Complex). Some 800 km from Montreal, Manic-5 is more accessible than James Bay (which is 1300 km from Montreal), so it's where most "hydroelectric tourists" like us head to learn about where electricity comes from and how it is produced.

We didn't think more than a handful of people would make the long trek to Manic-5. As an impressive example of "Brutalist" architecture—a style based on using raw materials and simple geometric forms—Manic-5 is worth seeing. Still, it takes more than two hours to get there from Baie-Comeau, which is itself a long day's drive from Montreal. There isn't much to see between Baie-Comeau and Manic-5. Highway 389 is a long winding, isolated drive through endless mountains, marshes and spruce forests. The only visual breaks are the imposing Manic-2 dam, a substation and road signs for the turnoffs to Manic-3 and Outardes-4, each of which takes another hour off the main road to get to.

So imagine our surprise when we arrived at Manic-5 to discover the visitor parking lot was already full. Our group, on the 1 pm tour, was the third of the day and it was so big our tour guides had to find an extra bus for it. There was even a third bus with foreign tourists arriving from Quebec City following us through the tour. The hydroelectric tourists we met included a handful of nostalgic former employees of Hydro-Québec and a few members of the group told us they had discovered a new passion for electricity after seeing Christine Beaulieu's docudrama play, *J'aime Hydro*, which was the surprise hit of Quebec's 2017-2018 theatre season and still continues to draw crowds. But most of the visitors were Quebeckers of different ages and origins who were just curious about hydroelectricity and wanted to learn more about Quebec's fabled Golden Age of dam building.

Fig. 0-B: Tourist bus at the foot of Manic-5's Daniel-Johnson Dam. The dam is as high as the tower of Montreal's Olympic Stadium.

With its 13 immense arches stretched over 1.3 km of the Manicouagan Valley, Manic-5's Daniel-Johnson Dam is the most spectacular of Hydro-Québec's 63 hydroelectric power stations. Gazing at the soaring vaults from both above and below, and strolling through the interior of the dam structure makes for a mind-boggling experience. When we finally saw "La Manic," as Quebeckers call it, we weren't surprised so many people drive so many hours just to lay eyes on it. The dam is a powerful symbol of ambition for Quebeckers. It's also fascinating to learn about the mysteries of the invisible "fluid" called electricity, and realize what an incredible feat heating homes and powering factories across the province really is.

"La Manic" struck the collective imagination of Quebeckers years before the dam opened in 1970. Guided tours of the site started in 1964: visitors at the time could only see a mock-up of the finished dam and observe the work site from a distance. Renowned Quebec singer-songwriters Georges Dor and Félix Leclerc started recording songs about Manic-5 in 1965. The celebrated Belgian cartoonist Hergé (Georges Prosper Remi) visited the station while it was still under construction and drew sketches of his famous characters Tintin and Snowy beside the future Daniel-Johnson Dam (the drawings are stored in Hydro-Québec's archives but the company unfortunately never obtained the rights to display them). "La Manic" even inspired a novel in best-selling Belgian novelist Henri Vernes's Bob Morane adventure series (*Terreur à la Manicouagan*, number 71). The protagonist of Vernes's story is pitted against the fearsome villains Miss Ylang-Ylang and Roman Orgonetz who are trying to blow up the dam. During Expo '67, visitors to the Industry of Quebec pavilion could watch live footage of the dam and power station being built. At the beginning of the 1970s, Quebec even produced a sports car, the "Manic GT," named after the dam: roughly a hundred models were built in a factory in Granby before it shut down in 1971.

Judging by the displays at Manic-5's visitors centre, it's fair to say that most hydroelectric tourists are interested in history. The Golden Age of dam construction was a key element of Quebec's Quiet Revolution in the 1960s and 1970s, when Quebec society modernized rapidly. As for the future, it's the trip to Manic-5 along Highway 389 that provides a window onto the challenges in store for Hydro-Québec. That is what interested us when we began writing this book.

At Manic-5, Hydro-Québec built a second power station next to the first one in order to meet Quebeckers' insatiable need for electric energy. Yet one can also see water from the dams being emptied into the dam's spillways. This is due to the fact that at the moment, Hydro-Québec actually has a surplus of energy. The company either has to find a way to use that energy, or it will be wasted—so Hydro-Québec is earnestly looking for new places to sell it. Not as simple as it sounds.

The great paradox of Hydro-Québec in 2019 is that it has too much energy but not enough power. Grasping the difference between power and energy is essential to understanding the challenges Hydro-Québec will be facing. There are signs of this dilemma all along Highway 389. Halfway between Manic-5 and Baie-Comeau, Hydro-Québec built a new power line starting at the Micoua substation at kilometre 94. It's one of the measures the company has taken in the face of stagnating electricity requirements in the last 10 years. Hydro-Québec built the line as a way to release the surplus energy of the dam, to send the energy somewhere it can be used. Finally, the gigantic Churchill Falls generating station in Labrador, stands at the very end of Highway 389. Hydro-Québec has been buying almost all the energy this station has produced since 1976. But the contract Hydro-Québec signed with Newfoundland-Labrador is set to expire in 2041. And when that happens, Hydro-Québec will have to find a way to replace the 5000 megawatts of power that the company has been buying at bargain basement prices for over 40 years now. Hydro-Québec's directors are already searching for a solution.

This sounds paradoxical, even contradictory. Does Hydro-Québec have too much electricity? Or not enough? In fact, it boils down to one fact, essential to understanding the future of electricity: *energy* and *power* are not the same thing, and managing them poses entirely different problems to Hydro-Québec. Hydro-Québec has *too much* energy (at the moment). Yet at the very same time, it does not have *enough* power to ensure a supply to Quebeckers in the future. Understanding this is essential to grasping the challenges in Hydro-Québec's future.

And speaking of the future, it's catching up with Hydro-Québec in other ways as well. In a generation's time, the state-owned enterprise will be functioning in an entirely new technological, social and economic environment. It must start evolving to meet those challenges, now. Changes will also come from the behaviour of its millions of customers in Quebec, the United

States, Ontario and elsewhere. The recent outcry of residents in Saint-Adolphe-d'Howard protesting the construction of a high-voltage line through their small town in the Laurentians region is just one sign of new consumer attitudes. And of course, the future of Hydro-Québec depends on actions taken by both the Quebec government and the governments of its neighbours, in particular Newfoundland and Ontario.

Our research has shown that Hydro-Québec is not frozen in place like the proverbial deer in the headlights (we thought a better image would be "a caribou on the ice of the Caniapiscau River") in the face of these challenges. On the contrary, the company is preparing for the future by developing new markets, new services and new means of communication which, together, will allow it to remain profitable and relevant. When we interviewed Quebec Premier François Legault in January 2019 about Hydro-Québec's export projects, he referred to Hydro-Québec as an "underexploited jewel"[2] and told us he had great ambitions for it.

So it is with a resolutely futuristic perspective—forward-looking, but not idealistic—that we started writing about Hydro-Québec. No one at Hydro-Québec's headquarters has a crystal ball. By 2022, its managers will have to decide if they want to build a new dam by 2040. And they will have to base decisions on predictions of future needs—needs that are hard to determine amid radical changes in the energy industry. There are a lot of unanswered questions. What gains in efficiency can they expect to make in a generation? What will happen to Quebec's aluminum factories, which are among the biggest consumers of electricity in the province? Will the electric car live up to its promises? Will Quebec succeed in attracting massive electricity consuming businesses like data centres? How far will Quebec's neighbours go in their commitment to decarbonize their economies and push solar energy? Will Ontario abandon nuclear energy once and for all? Will Indigenous Peoples, and in particular the Innu, demand a nation-to-nation treaty over hydroelectric resources like the Cree peoples and Inuit did before them? All these unknowns, plus other trends still unfolding, promise to have a profound impact on the future direction of Hydro-Québec.

The future is also built on past decisions: Hydro-Québec's work has always been to make predictions on horizons of 20, 30 or 50 years. The reason Hydro-Québec is one of the best managed power systems on the continent today, earning the praise and admiration of the energy sector's leading companies, is that the company and government—and through them, Quebec society on the whole—have made more good calls than bad over the last 75 years. Some strategic decisions proved to be more than astute: they were audacious. One reason Hydro-Québec has earned a reputation for being well managed is that its leaders have learned from the company's mistakes—and they certainly have made mistakes.

There is nothing surprising about this when you step back and look at how well Quebeckers have managed the incredibly complex task of producing, transporting and distributing electrons across one of the biggest systems in the world.

Chapter One

A Tour de Force

In January 2010 we moved to Phoenix Arizona for six months to work on a book project. We wanted the experience to be a family adventure, so we bought an old RV and drove to Phoenix through the Midwestern United States and the Rocky Mountains. Once we got settled into life in a rented bungalow in Phoenix we used the RV for weekend excursions to southern Arizona, Mexico, California, the Grand Canyon and beyond.

With its steady supply of warm, sunny weather, Phoenix is a wonderful place to live in the winter, but everything changes at the end of April when the thermometer starts flirting with temperatures of 40°C. As the heat rose, Jean-Benoît got into the habit of plugging in the Winnebago the night before we set out on trips to make sure the fridge would be cold when we left the next day. Then one May evening, after a particularly torrid day, we couldn't get the fridge in the RV to switch from propane into electrical mode. Jean-Benoît pulled out his voltmeter, an instrument every RV owner has to check the voltage in campgrounds, which is unpredictable. That night the voltage of our bungalow was only 100 when it should have been 110 or 120 volts. Inside the house, we noticed the lighting was slightly more "yellow" than normal.

We were experiencing our first Phoenix "brownout." It's not an unusual occurrence in southern U.S. cities where power systems struggle to meet the high demand for electricity when air conditioners are running full blast during the hot summer. To avoid blackouts, electricity suppliers simply reduce the voltage of the whole system by using a technique called load shedding. Load shedding produces controlled brownouts, which as we would learn, are pretty common in Phoenix in the summer. Most of the time the only discernable effect are dim lights, but brownouts can sometimes mess up digital controls and electrical appliances, or even make motors run backwards. In fact, every year, load shedding burns tens of thousands of fridge motors, air conditioners and other electrical appliances in the U.S. Some of our neighbours had backup batteries to protect sensitive appliances. In short, when brownouts happen, people just work around them.

As Quebeckers, we had a hard time taking these power fluctuations in stride. Load shedding is practically unheard of in Quebec. While brownouts are a part of daily life in places like Phoenix, Montrealers and other Quebeckers would never put up with power fluctuations that regularly brown our light or fry the motors of our driers and toasters. Hydro-Québec's customer services lines would be jammed. Load shedding is practically unheard of in Quebec.

The experience got us thinking just how efficient and reliable our energy supplier actually is.

Hydro-Québec has one of the biggest power systems on the continent, with the most reliable service. Taking this service for granted, Quebeckers are blissfully ignorant of the incredible technical mastery required to produce, transmit and distribute electricity across their huge territory. Like stores that sell milk, meat, eggs or fish, Hydro-Québec supplies a kind of "juice" that is a basic necessity. The difference is that electrical "juice" has to be delivered in precisely the right quantity and quality at all times: voltage, frequency and wattage of electricity must be absolutely balanced. In ensuring this, managers of the grid actually perform a job similar to that of sound engineers in a recording studio. Specialized audio technicians do more than just listen to volume.

Fig. 1-A: The System Control Centre (SCC), the brains of Hydro-Québec, located somewhere in Montreal. The location is kept secret for security reasons.

They have to be sensitive to tone, rhythm and harmonics and alert to echoes, feedback or other interferences. In the case of electrical waves, "interference" can end up damaging and breaking equipment, or shutting the entire system down.

When Hydro-Québec was founded in 1944, most Quebeckers only used electricity to power a few light bulbs and a couple of stove burners, at the most. The quality of the electrical wave was not much of an issue in daily life. The situation is entirely different today. We plug millions of computers, alarm systems and electronic control systems into electrical outlets without giving it a second thought. On top of that, electricity powers aluminum smelters, factories, stores and gigantic server farms across Quebec. All of them require a perfectly calibrated electrical wave to function.

The reliability of Hydro-Québec's system is all the more impressive considering that Quebec's territory is larger than the entire southwest of the United States, including California. To supply this vast area, Hydro-Québec has a virtual army of 20,000 employees ready to work in all weather conditions, in even the remotest corners of the province. Hydro-Québec's system consists of 352 turbines and 63 generating stations, about a billion switches, power outlets, light bulbs and appliances of some 4.3 million customers, and enough power lines to wrap around the planet six and a half times. And the whole thing must be perfectly balanced, a phenomenon everyone takes for granted.

A Cure for Magical Thinking

To be fair, few of us have more than a basic grasp of how electricity actually works in the first place.

When an amazed public saw Thomas Edison light an electric bulb for the first time in 1879, the effect was so magical people talked of an "electricity fairy." But electricity is not magic: it is a phenomenon fully explained by physics. It's also one that requires a high level of technical mastery to produce and manage.

So, what is electricity exactly? It runs through wires, yet we never "see" it—only its effects, when lights come on, images move across screens or appliances start working. At most, plugging an appliance into an outlet will occasionally produce a minuscule electric arc (when electricity jumps from one conductive element to another), which is essentially a tiny bolt of lightning. Otherwise, we "see" electricity in its static form, when it makes hair stand on end, or makes a balloon stick to a wall. It can also create a magnetic field that moves the needle of a compass or makes a pacemaker malfunction. Yet while electricity can be powerful enough to enable a crane to lift 100 tonnes of rock, it doesn't weigh a thing. A charged battery is no heavier than an empty one.

Electricity is a phenomenon of physics—its first discoverers described it as a "fluid" and a "transfer agent."[1] More specifically, electricity is electronic: the effect of an electron jumping the orbit

Fig. 1-B: The difference between power and energy. The water jet represents the intensity, or power, expressed in watts, kilowatts or megawatts. The water that fills the bathtub is energy, expressed in watt-hours or kilowatt-hours.

of its nucleus in order to move toward another atom. This move sets off a chain reaction that moves one atom to another. The result is an electrical current.

A number of images can be used to illustrate this phenomenon. The electricity in a house, for example, is analogous to pipes in a plumbing system (Fig. 1-B): the pushing force of the water (or pressure) that allows it to move one way or the other, corresponds to volts in electricity. Volts are what makes the current (measured in amperes) move through the power line. But the degree of force of the water leaving the pipes, the intensity of the flow, is analogous to electrical power, which is expressed in watts. The quantity of water that flows out and fills a bathtub in a given time, is energy, expressed in kilowatt-hours.[2]

But there are limits to this analogy, because unlike water, electricity doesn't weigh anything and it moves at something close to the speed of light. Electrical current is, in fact, a wave. The electricity in our outlets and lamps is undulating, like waves in the sea or a backyard swing set, except much, much faster: at 60 cycles (up and down, up and down) per second. And while electric current undulates, it can't just swing randomly. For the system to work, the system managers have to precisely coordinate each of Quebec's 352 turbines and 63 generating stations. In the alternators at the Beauharnois generating station near Montreal, or at La Grande-4 in James Bay, Rapide-des-Quinze in Abitibi-Témiscamingue or La Romaine-4 in the Côte-Nord, the magnets in the rotors have to pass in front of the conductive bars in the stator at *exactly* the same time. Across Quebec, the electrons over 260,000 km of power lines have to undulate in perfect synchronicity, all the way to the hairdryers in bathrooms. It's as if eight million Quebeckers in their offices and homes were singing the same song all day and all night, in the same key, in perfect unison.

Without this little miracle of synchronization, there wouldn't be a power system, period.

The best image to illustrate the force of electricity is probably Newton's cradle. Most of us have seen one of these little metal frames with five suspended balls on an office desk somewhere. When the first ball is lifted, it falls and hits the four that are immobile, but the three in the middle stay perfectly still while the shock wave makes the fifth ball fly up. Electricity works the same way. When one electron is subjected to a force that moves it, the shock wave moves through a chain of electrons and exits at the other end with the same force. Except the first electron has to be pushed with enough force to light streets, power factories and operate a myriad of home appliances from microwave ovens and computers to radios, food processors, lamps and printers. And the length of the "chain" can be up to 1500 km away—the distance from the Robert-Bourassa generating station near James Bay to Laval.

While the plumbing system, swing set and Newton's cradle analogies are helpful simplifications, the phenomenon of electricity actually requires hundreds of pages to describe and mathematically formulate. Before Samuel Morse commercialized the first electric telegraph in the 1840s, electricity was mainly a curiosity that was used to impress crowds at exhibitions. It took a small handful of enlightened scientists who carried out rudimentary experiments (and made deductions worthy of Hercule Poirot) to subsequently describe the process. Following Thomas Edison's invention of the light bulb in 1879 and the creation of the first electric systems at the beginning of the following decades, an avalanche of discoveries, inventions and technical breakthroughs would turn electricity into one of the foundations of modern life.

The Strength of Water

Jean-Benoît has flown over the Robert-Bourassa generating station twice by helicopter, in 1998 and 2018. Each time the experience was breathtaking. Robert-Bourassa, 1500 kilometres north of Montreal, which was previously known by its abbreviation LG2, is the second of five hydroelectric power stations on the La Grande River that empties into James Bay, a gulf of Hudson Bay. The scale of this gigantic electricity "factory" can only be appreciated from the air. The endless reservoir, three times the size of Lac Saint-Jean, collects water from a territory the size of France. The spillway for the reservoir, held back by an immense dam of stone and rock, wasn't dubbed the Stairway of Giants for nothing. For 45 years, 100,000 workers and engineers worked building several hundred dams and dykes necessary to operate 11 hydroelectric generating stations. It was an enormous job for which 1500 km of roads, dozens of bridges, 5 airports, and 5 workers' camps, each the size of a small city, had to be built.

It wasn't the aerial view of the station that impressed Jean-Benoît the most, though, it was the view by bus as he approached

Fig 1-C The Robert-Bourassa Dam and a corner of the reservoir, which is three times the size of Lac Saint-Jean. Left, the floodgates of the gigantic spillway, dubbed the Stairway of Giants. The generating station, located at the bottom of the valley, is invisible: it can only be reached through a kilometre-long tunnel.

the power station at the bottom of the valley. Only when the vehicle passed through a set of immense doors leading into the underground tunnels did Jean-Benoît grasp the incredible scale of the work it took to transform this valley into a wall as high as Montreal's Olympic Stadium tower. The wall, raised through extraordinary effort to hold back an enormous mass of water, is what really represents the strength of the hydroelectric complex that generates 7772 megawatts of electricity. The dam feeds two generating stations, Robert-Bourassa and LG2-A, forming the biggest hydroelectric complex north of the Rio Grande.

The visitor bus passes through an enormous doorway where, two generations ago, 100-tonne "Tonka" trucks drove in and out as they gradually emptied the guts of the mountain. The bus then travels down a kilometre-long tunnel to a sort of roundabout dug into the granite. Visitors enter the immense underground gallery on foot. It's the biggest underground power generating station in the world, so vast that employees use bicycles or even small cars to get around the facility. Sixteen powerful turbines are anchored in the bedrock, making the ground vibrate. Above them, a torrent of water more massive than Niagara Falls pours into 16 penstocks (enclosed pipes that deliver water to hydro turbines) and 270 tonnes of water per second (enough to fill an Olympic-sized pool every nine seconds) hit the blades of the huge turbines like hammers striking an anvil.

It's not, properly speaking, the turbine that produces electricity, but a "generating unit": the turbine activates an alternator mounted on top of it. This unit, which can weigh up to 500 tonnes, could pass for the electrical motor of a giant toy (Fig. 1-D); the two are almost identical. In an electric toy, the battery energizes a coil of wire that induces a magnetic field. The strength of this magnetic

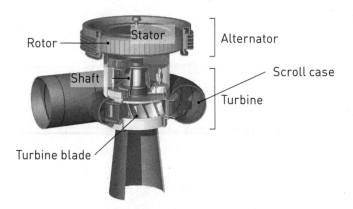

Fig. 1-D: Water enters the turbine through a pipe called a scroll case. It then hits the blades of the turbine, which sets the shaft in motion, activating the rotor in the alternator. The rotor, clad with magnets, excites the electrons on the copper plates of the stator. The magnets are what produce electricity.

field pushes the magnets attached to a rotor that, in turn, activates the wheels. The turbine of the dam executes the same task in reverse. Instead of wheels, there is a turbine. The mechanical force of the water activates the turbine that turns a rotor clad with 54 magnetic poles. When the magnets of the rotor move in front of the 980 copper electrical conductors of the stator, their magnetic field excites the electrons and induces an electrical current. The electric current then flows to Montreal.

The James Bay Project, whose construction began in 1973, reached its full size in 2012. It now takes 900 employees—electricians, engineers, electrical mechanics, security agents, controllers—to keep its 12 generating stations running properly. For operations like these in remote regions of Quebec, most of the employees work what they call an "eight-six": eight work days up north with six days off at home in the south. A few employees live up north in places like the village of Radisson, population 270, where there is a hotel and a few bars and corner stores. But most employees shuttle back and forth to the north, where they live in dormitories with a cafeteria and (for the lucky ones) a gym and social club.

Power dams might look too solid to budge, but they actually have to be constantly monitored. Water, ice, water runoff or seismic tremors can damage them. Hydro-Québec uses sophisticated instruments to detect the smallest of movements in its dams. When we visited the Beauharnois station, which transforms the energy of the St. Lawrence River as it arrives in Montreal, our guide pointed to a series of hydraulic jacks arranged horizontally in the dam. A technician comes to tighten them a bit each month, he explained. Since Beauharnois began operating in 1932, a chemical reaction that causes the concrete to expand has pushed the dam forward a total of 8 cm. There's no real danger. The movement is closely monitored. But every year since 1995, the generating station has to be pulled back a few millimetres to put it back on its original base.

All of which is to say, when it comes to managing a network of 600 dykes and dams like Hydro-Québec's, there's no improvising. Everything is calculated, planned and monitored by the 2900 employees of Hydro-Québec Production who oil the machines, inspect access to the power stations, calibrate measuring apparatus and much more. Technicians measure the amount of snow that falls before each thaw. Mathematicians calculate the water level of the reservoirs. Glaciologists prevent underwater ice from forming and jamming the turbines. Biologists determine the best way to use the dams so the spawning period for different species of fish isn't threatened. Meteorologists predict the floods that might occur upstream or downstream from a dam. The power station managers decide whether or not to release water from the dams.

The Control Tower

Don't go looking for Hydro-Québec's System Control Centre (SCC) among the skyscrapers of downtown Montreal: the centre's location is kept secret for national and continental security reasons. That shouldn't come as a surprise. Like air traffic controllers in an airport that will never shut down, three system operators work around the clock in the centre to keep Hydro-Québec's system "flying." They are responsible for the comfort of Hydro-Québec 4.3 million account holders (and their families) as well as millions of other customers in neighbouring systems that are connected to Hydro-Québec's.

The SCC is literally an isolated, windowless bunker. With its own autonomous air conditioning system, water and electrical supply, the centre was designed to resist sabotage or espionage attempts, terrorist attacks and computer hacking. The security system is similar to that of an airport. To get to the SCC, we walked down a narrow hallway lined with surveillance cameras to a private elevator. On exiting, we were greeted by an SCC security guard who was sitting behind a bullet-proof window. He asked to see our driver's licenses, scrupulously looked them over and then handed us a security badge. Then the SCC's manager personally opened each of the three doors we had to pass through to access the control room. And even when we got to it, we couldn't actually enter the room: we had to observe operations through a window.

On the back walls of the dimly lit SCC we could see a huge synoptic table arranged in a half circle with gently flickering lights. The table schematizes Hydro-Québec's 63 generating stations, 553 substations and 34,500 km of high-voltage power lines (see map, p. 13). The back wall of the SCC brings to mind the monitoring system of an intensive care unit. And that's not too far. The wall monitors the vital signs of Quebec's enormous power system. Digital dials in red, green and yellow indicate the level of power being produced at each generating station, the number of turbines in service at any moment, and the load; the data is updated every three seconds by way of a fibre-optic and carrier wave telecommunications system. Above this data a digital clock displays Hydro-Québec's ultimate reference number, 60, which indicates that the system is functioning at the proper frequency of 60 cycles per second.

The three system operators in the SCC sat calmly at their desks during the half hour we watched them work. From time to time, one of them would pick up a phone to confirm an operation, but most of the time, their eyes didn't leave the instruments and weather reports. "The goal is to be able to deal with any damage that might come about as a result of weather changing or equipment breaking. The challenges vary with the seasons," explained Francis Monette, Manager, System Control and Programming at Hydro-Québec. In addition to daily peaks in demand, the system has to manage large annual variations in power demand. At 5 a.m. on June 24 (St-Jean-Baptiste Day, Quebec's national holiday), when no one is working, demand can bottom out at less than 14,000 megawatts, but on December 28, 2017, it momentarily hit a high of 38,420 megawatts at exactly 4:28 p.m. "During winter peaks in demand, things here are buzzing, the entire system is being used and our work is to keep everything in balance," says Francis Monette. "But strangely, it's in summer, when demand is lower, that things get precarious. We do repairs on lines and turbines in the low season, so there are fewer pathways available for energy to pass through the system."

The truth of the matter is that the system is never completely protected from the kind of damage that can set off a cascade effect that will eventually trigger a blackout. That's exactly what happened on July 3, 2013, during peak load period, when a forest fire threatened power lines at James Bay. In cases like this, the fire itself is not what poses a risk to the high-voltage lines: it's the smoke, which ionizes the air and makes it as conductive as metal, creating sparks that lead to short circuits. The inevitable happened on July 3. The fire got too close to the line and the dense smoke provoked electric arcs between the line and the smoke. To protect itself, the system automatically flipped its breakers, cutting the current in the line. Since there is no alternate line that can be used as a backup to supply power, other mechanisms automatically cut the electricity supply to Hydro-Québec's 1800 industrial and commercial customers. That led to an outage in Montreal's Métro system and at La Ronde amusement park, which then had to be evacuated. It took more than an hour to fix the problem by starting up other generating stations and buying electricity from neighbouring systems. "We're making sure that won't happen again," says Francis Monette.

Over the years, researchers at Hydro-Québec have put a series of automatic controls in place to solve virtually any problem in the system within milliseconds. Hydro-Québec has procedures for all "normal" problems—in principle. But sometimes the computers can't solve the problem. That's exactly what happened on December 4, 2014, when a pilot with a grudge against Hydro-Québec carried out an air attack on one of the high-voltage lines that crosses the Laurentians region north of Montreal. Details about the attack have been kept secret for national security reasons. Flying his single engine propeller plane, the pilot managed to interrupt the electrical current in mid-flight. His act of sabotage left 200,000 customers in the dark and almost caused a system-wide outage—something that hadn't happened in Quebec since 1989. System operators at the SCC only had a few seconds to recover. The paper mill Kruger agreed to interrupt operations *in extremis* to save the system and the power operators managed to make up for the power shortage by getting help from neighbouring systems, though at a hefty price. "It cost us 28 million dollars," says Francis Monette. The system was in a precarious state for a whole day.

There are 24 system operators altogether at the SCC. They monitor the system 24 hours a day in teams of three. The transmission system operator, who sits in the middle desk, is like an orchestra conductor, making sure the system as a whole is stable. Most of the time the system works on something like automatic pilot: computers manage most of the settings in real time. But the transmission system operator has to know what is going on every second of the day in order to be ready to take over the controls if necessary. Turbines starting or stopping, weather changing, variations in demand or other elements like solar wind or lightning strikes—which happen up to a million times per year, or 8000 per hour during a storm front—are all factors that affect the voltage of the power system.

The interconnections system operator is in charge of the 15 interconnections Hydro-Québec has with its neighbouring power systems. These neighbouring systems either receive or deliver specific quantities of energy at fixed times. The interconnections operator is busy: Hydro-Québec supplies 25 percent of Vermont's energy, 15 percent of New England's and New Brunswick's, 5 percent of New York State's and 4 percent of the electricity consumption in Ontario. These are Hydro-Québec's main markets but it has also sold megawatts as far away as North Carolina, Tennessee and Indiana.

The generation system operator's job is to make sure hydro generation meets demand. What comes into play here is not so much the law of the market or economics, but simply the laws of physics. Specifically, Kirchhoff's laws, named in honour of the 19th century German physicist who formulated the rules that govern the functioning of electrical circuits, whether they are powering a toy or a large system. The gist of the laws is that what enters an electrical circuit must equal what leaves it. Energy supply has to be in constant balance with demand. There can be no surplus left over afterward, but no shortage either. If the amount of electricity consumed exceeds the quantity produced, a circuit breaker will trip, provoking a local outage or a series of cascading outages that can escalate into a general power outage. But if the quantity of electricity generated exceeds consumption, the system will shake like a pipe with too much water pressure, which risks damaging equipment, and, once again, raises the likelihood of a general power outage. The generation operator has to be able to determine, at any given moment, whether Hydro-Québec is in a position to produce the amount of electricity required, which varies all the time, or whether it will have to buy some, or be in a position to sell a surplus to outside markets. The generation operator manages this energy "balance sheet" and either starts or stops turbines or generating stations to keep supply and demand in balance.

Behind the 24 system operators at the SSC, another 150 employees study weather variations, keep track of how much generating capacity is available, from which stations, and measure the volume of water available in each reservoir. "For us, predictions change every 10 minutes," says Francis Monette. Hydro-Québec has economists and statisticians studying power demand all day long. There are five meteorologists analyzing the most minute details of weather data. The questions literally never stop. The analysts have to predict things like whether the thermometer will really fall

Fig 1-E: Twisted electrical towers illustrate the gravity of the ice storm that hit Quebec in January 1998 and destroyed a large part of the electrical system in the southern part of the province. To make sure history wouldn't repeat itself, Hydro-Québec rethought the system from top to bottom: designing better towers, making the lines more robust, installing high-performance ice detectors and improving civil protection measures.

to -20 °C at 4 a.m., or whether it will hit its low at 8 a.m., right in the middle of the peak morning load period. What will the wind speed be? Will there be clouds or not? "Twenty years ago, the SCC really worked with its nose to the ground, reacting to every single weather change," explains Francis Monette. "Today we have managed to boost system reliability so much that we can work almost entirely based on weather forecasts, in other words, more based on what's coming than what's actually happening."

In popular thinking, Hydro-Québec is associated with its spectacular power dams. Yet the backbone of the system is actually the 34,500 km of high-voltage power lines controlled by the SCC, which are linked to faraway energy generation centres. Unlike power dams, with their astonishing quantities of water pouring in and out, the work of a properly working power line wouldn't strike anyone as impressive at a glance. Sure, you sometimes hear a high-voltage power line buzzing or crackling in wet weather, but most people don't have a clue what's really going on inside. The 533 transformer stations where voltage in the wires is increased and lowered, aren't much more impressive, in appearance. The only time anything remarkable happens is when a line gets struck by lightning or weighed down by freezing rain (Fig. 1-E). Unless weather conditions damage them, good lines last 80 years. During that time nothing happens to them—at least nothing visible.

Hydro-Québec devotes huge resources to making sure its lines and transformer stations keep quietly doing their job: in 2017, the company spent $2 billion on line maintenance, which is over half its overall investment budget of $3.8 billion. The 3000 employees at Hydro-Québec's TransÉnergie division head out by truck, helicopter, snowmobile, on lifts and on foot to inspect pylons, lines, transformers and circuit breakers—when they aren't replacing them. That doesn't include the employees who maintain the other system Hydro-Québec's put in place to control the system: its telecommunications network.

Fire Fighters

For simplicity's sake, Hydro-Québec is often referred to as a "system." In fact, the company is a combination of three systems. We have described the high-voltage system with its 34,500 km of lines that transmit electric power between 69,000 and 735,000 volts— or 7000 times the voltage of a toaster. The high-voltage system is managed by TransÉnergie, from the SCC. But Hydro-Québec also runs two other, much larger systems. The first is the low-voltage system: 106,000 kilometres of wires that carry between 120 and 600 volts through the streets of Quebec and are linked to electricity meters in houses. Between the high and low-voltage systems, there is a third system of lines of between 4000 and 34,000 volts. This medium-voltage grid basically works like an irrigation system: it carries electricity from the high to low-voltage systems to make it easier to distribute.

Hydro-Québec Distribution has a very complicated job. Besides managing the low and medium-voltage systems, a total of 224,000 km of lines, the division is responsible for 2 billion electric poles, 680,000 overhead transformers, another 34,000 underground vaults and 4.3 million electric meters. So it should come as no surprise that with 5400 employees, Hydro-Québec Distribution has almost as much personnel as Hydro-Québec Production and TransÉnergie combined. In addition to managing the hardware required to carry electrical current to houses, stores, factories, hospitals and schools, Hydro-Québec Distribution also has to manage people. Customer Service has 800 employees. For the last 25 years, computers have been generating invoices that are sent by mail, but it takes real bodies to answer customers' questions and complaints and manage connections and disconnections with a smile even when customers are unhappy.

This work is so complicated that Hydro-Québec Distribution has a whole separate brain to manage things, the Distribution activities management centre (Centre de gestion des activités de distribution, CGAD). Located in a building near Highway 40 in Montreal, the CGAD manages and coordinates the activities of its five administrative territories. As with the SCC, whose location is kept secret, Hydro-Québec prefers not to advertise the exact location of the headquarters in each territory, which correspond roughly to the Bas-Saint-Laurent, Quebec's capital (Quebec City), Laurentian and Montérégie regions, and Montreal.

The CGAD's nervous system is the Distribution network coordination centre (Centre de coordination du réseau de distribution, CCRD), a vast room whose walls are covered in blinking screens displaying data about the state of the network, along with weather charts showing wind direction and cloud formation. Contrary to the SCC, which is literally cut off from the world, the CCRD has open areas for impromptu ad hoc committee meetings. Right behind the CCRD, a crisis centre is visible through a glass wall; the rest of the walls are, again, covered with blinking screens. This is where managers and members of the communications team meet whenever there's a serious power outage, like the one caused by tornadoes in the Gatineau region in September 2018, or the violent storm that struck the Magdalen Islands a month later.

"There is always something going on, somewhere, a branch falling on wire, a transformer giving up the ghost, or a car running into an electrical post. And when it's not the weather, something else happened, or someone did something," says Patrice Richard, Manager, System Operation Activities at Hydro-Québec.

One factor complicates management of the distribution system: the fact that the same people handle routine network maintenance and emergencies. It's as if firefighters had to put out fires and do maintenance work on all the houses in the system at the same time. Yet that's the daily lot of Hydro-Québec's 1500 line and cable workers, and of the logistics and planning teams who manage them. They do whatever they can to solve the smallest outage, then must head back to their routine maintenance work pick up they left off.

Though he's like the "fire chief" in the system, Patrice Richard also manages customer perception. This was in fact the original motive behind the creation of the CGAD in 2013. "Too often we learn about a problem in the system from social media, not from our own staff," says Patrice Richard, who considers "coordination" the key word in the name of the CCRD. "We're here to monitor the whole system. Each of our territorial divisions carries out their normal activities separately. We only get involved when there are delays in work or an emergency that a certain territory doesn't have the capacity to handle on its own."

The CCRD's other job is to make sure TransÉnergie, Hydro-Québec's transmission system, delivers enough electricity to the distribution grid to meet demand, so they can prevent local outages and brownouts. "Our engineers can take energy from one territory and allocate it to another so we can avoid asking TransÉnergie for more, but we have to know when it's the right time to do that," explains Patrice Richard. The distribution system also has to adapt to the population, which requires almost constant re-planning. The sudden rise in population in the Laurentians region 10 years ago, for instance became a source of headaches for Hydro-Québec, and the same phenomenon is going on in the Montérégie region south of Montreal.

"In theory," says Patrice Richard, "every substation needs three transformers to function, and TransÉnergie puts a fourth in as a backup. But sometimes there are only two that work when we need three. Hydro-Québec is a very big machine and not the easiest to keep running some days."

Chapter Two

Technical Prowess

The first time we saw the giant black cube in the middle of a field in Varennes, just off Highway 30 on Montreal's south shore, it made us think of the film *2001: A Space Odyssey*. This eerie windowless block made entirely out of metal is actually an enclosure built to block electromagnetic fields. In physics jargon, it's called a Faraday cage. The contrast between the gloomy exterior and the building's contents struck us. When we stepped inside, we thought we were either looking at modernist sculptures from Expo '67, or on the set of an *Austin Powers* movie. Filled with an amazing array of almost beautiful measuring devices of fantastic shapes and sizes, the building felt like a futuristic playground.

It is actually a playground—for scientists. The laboratory in Varennes is actually part of the campus of the Hydro-Québec's research institute (Institut de recherche d'Hydro-Québec, IREQ), Hydro-Québec's scientific research arm.[1] The instruments are not toys, but extremely sensitive devices developed to obtain precise measurements of different electric and magnetic phenomena. Since IREQ opened in 1970, Hydro-Québec has been carrying out research vital to its future here, including thousands of projects and experiments on electricity generation, transmission and distribution.

Wandering around this electricity research heaven, we started to feel like we were on the set of a James Bond film, waiting for the

Fig. 2-A: The IREQ campus in Varennes. Hydro-Québec is the only electric utility on the continent that has its own research centre. In the United States, it's the federal government that carries out this kind of research.

eccentric Q to pop out and show us the latest gadget he'd dreamed up for British secret service agents. In a way, we weren't too far off: we actually did see prototypes of robots—flying, underwater and rolling—that Hydro-Québec is developing to inspect its networks. We also saw batteries the size of a tractor trailer. Hydro-Québec designed an electric car wheel at IREQ in the 1990s. In the late 1980s, it built a Tokomak Fusion Test Reactor here to carry out experiments in nuclear fusion.

Of all the instruments, it was perhaps Hydro-Québec's brand new line-inspecting robot, the Line Ranger, that impressed us the most. Among the latest crop of robots the company has designed and built, the Line Ranger looks like a small spatial module (this time, we thought we were in a George Lucas film). Equipped with a camera, a radiographic X-ray system, a mechanical arm and various sensors, the robot literally rolls along power lines on its own, at the rate of over 20 km per day, detecting the smallest breaks or repairing strands of wire damaged by lightning, all without interrupting electric supply to customers.

The Line Ranger is the latest milestone in Hydro-Québec's effort to develop and maintain what is really the backbone of its grid: the 12,000 km of very high-voltage lines that account for a third of the 34,500-km high-voltage system. The term "backbone" is not an overstatement: though seldom recognized as such, the 735,000 volt (735 kV) line is a fundamental innovation and in many ways the foundation for Hydro-Québec's entire grid. These lines are what makes it possible for Hydro-Québec to transmit very large amounts of electricity over extremely long distances, from generating stations as far away as James Bay, Quebec's Côte-Nord (North Shore of the St. Lawrence) and Labrador. Without this equipment, Hydro-Québec would never have become the largest energy supplier on the continent, and certainly not the greenest. Quebeckers would simply never have been able to take advantage of the hydroelectric potential of their territory. In short, the entire architecture of Quebec's hydroelectric system was made possible by the development of the 735-kV transmission lines.

The technology was invented in the 1960s, and to this day, remains Hydro-Québec's greatest technical accomplishment. The 735-kV line is also one of the greatest inventions ever produced in Quebec, up there with snowmobiles, snow blowers, telephone receivers, pig vaccines, the air-free baby bottle and the Wonderbra. Quebec's Order of Professional Technologists even named the 735-kV line the province's "technological innovation of the 20th century." Hydro-Québec's international reputation was built on the 735-kV line, even more than on its powerful dams. Hydro-Québec's research institute IREQ is, itself, a by-product of the effort to build the 735-kV line: to make it happen, Hydro-Québec had to scour the planet to find the best research minds.

Creating the 735-kV line was more than a technological feat. It was also a societal choice that would determine Québec's future as an energy producer. It's because of this technology that Hydro-Québec was able to build dams in faraway regions like James Bay, Côte-Nord and Churchill Falls, which together account for three quarters of Quebec's entire hydroelectric capacity. If not for the 735-kV line, instead of having one of the world's largest electrical systems, the province would have scattered conventional thermic or nuclear power stations along the St. Lawrence River, resulting in much more expensive and far less green electricity.

Instead, the 735-kV line turned Quebec into an oasis of renewable energy among other grids in the region, most of which are fuelled by coal, oil, natural gas or uranium. To be exact, 99.8 percent of the electricity produced in Quebec comes from renewable sources. Electricity accounts for 47 percent of Quebeckers' overall energy consumption, including what's used in transport and industry. It's a percentage of use of renewable resources that Ontario and California only dream of achieving in a generation or two. In fact, because the 735-kV line made it possible for Quebeckers to heat their houses and power their factories, the transition to renewable energy actually happened in Quebec in the 1970s. To top it off, Quebeckers did this while paying the lowest electricity rates on the North American continent—once again, thanks to the 735-kV line. If Hydro-Québec allows Quebec's neighbours to make the transition toward renewable energy, it will be largely thanks to the mastery of 735-kV transmission.

It's hard to overstate the contribution the 735-kV technology made to Quebec's hydroelectric grid. The feat was made possible by one brilliant decision, and a long series of technical breakthroughs dating back to when Elvis Presley was "electrifying" the youth of the Western world.

The Transmission Challenge

At the time Hydro-Québec was created, in 1944, it only operated four hydroelectric generating stations: one in Beauharnois, west

Fig. 2-B: The 735-kV line, one of the most important Canadian inventions of the 20th century. Thanks to this technology, Quebec has a 50-year lead over California and Ontario in the transition to renewable energy sources.

of Montreal; one in Rivières-des-Prairies on the eastern end of the island of Montreal; one on the Richelieu River in Chambly; and one in Les Cèdres, west of Montreal near Valleyfield. Like everywhere else in the Western world, energy demand was growing 7 percent per year in post-war Quebec, and doubling every 10 years. This rapid increase in demand forced electric utilities to launch new projects at a furious pace. From 1944 to 1962, Hydro-Québec multiplied its generating capacity sixfold, from 600 to 3700 megawatts.[2]

In 1950, Quebec still had a long way to go. A year before, Hydro-Québec set out to double, then in 1956, triple the output of the Beauharnois station, which harnesses the power of the St. Lawrence River as it arrives in Montreal. Classified as a national heritage site, Beauharnois appears a bit old-fashioned today, with its Art Deco architecture. When it was built, it was considered a technical tour de force, its creators having successfully diverted the St. Lawrence River by digging a 12-km headrace canal. Built in three stages between 1929 and 1961, with a kilometre-long turbine hall with 36 turbines producing 2000 megawatts of power, the Beauharnois station was the biggest generating station in the world at the time.

Though it only contributes 5 percent today, until the middle of the 1960s the Beauharnois station accounted for more than half of Hydro-Québec's overall electricity production. Seeing the frenzied increase in power needs in post-war Quebec, Hydro-Québec anticipated in 1950 that the Beauharnois station would soon fail meet the province's demands. In 1953 the company decided to build two new stations on the Betsiamites River on Quebec's North Shore, between Tadoussac and Baie-Comeau: we know it today as the Bersimis complex.[3] This 2000-megawatt project some 650 kilometres northeast of Montreal was Hydro-Québec's first megaproject. In 1955, seeing the exponential growth in power demand, Hydro-Québec started studies on yet another megaproject on two rivers north of Baie-Comeau, the Manicouagan and the Rivière-aux-Outardes, with seven generating stations that would supply 6700 megawatts of capacity.[4]

The enormous projects on Quebec's North Shore all faced the same problem: how to transmit electricity hundreds of kilometres to their main markets.

Weightless though it is, electricity remains one of the most difficult energy forms to move over long distances. When the current increases, lines tend to overheat and sag, and part of the energy dissipates in the air as heat. Another thing that happens is that the air around the lines becomes ionized, producing small discharges. This makes a distinct crackling sound and more energy is lost. Or voltage flutters and the lines become unstable. The system can also turn into a sort of huge antenna that captures not only radio noise but also solar wind and geomagnetic fields.[5]

While the Beauharnois generating station is only 50 km away from its load centre in Montreal, the Bersimis complex is located 500 km from Montreal as the crow flies; Manic is 600 km away. The bigger and further away from a new hydroelectric station gets from its markets, the more and longer the lines required become simply to transmit electricity to them. That exponentially increases the cost of construction. And since the energy in the lines dissipates over the distance it's transmitted, the lines end up delivering less electricity to their destinations. If Hydro-Québec had stuck to the technology available at the time for its new projects and transmitted 2000 megawatts from the Bersimis-1 and -2 to Montreal using 120,000-volt lines, the St. Lawrence River would have been overrun by electric lines and towers.

Instead, in 1950, only six years after Hydro-Québec was created, its engineers wiped the slate clean and started thinking about how many volts a power line could potentially carry. The question (and answer) might well seem obvious today, but at the time, it was a major technical obstacle. The future of hydroelectric power in Quebec, and maybe the world, was at stake, and the answers were far from evident.

It was a recent graduate of the Polytechnique Montréal engineering school, Jean-Jacques Archambault, who came up with the solution that would allow Hydro-Québec to become one of the world's leading hydroelectricity generators. A man with a low-key, reserved personality, Archambault was blessed with an immense mathematical talent and above all, uncommon strength of character. "He was by all measures exceptional, elegant and cultivated, a man who loved teaching," recalls André Bolduc, former economist at Hydro-Québec and author of numerous books on the state-owned company. "Jean-Jacques could motivate teams. I watched people push themselves to their limits for him."

In 1950, when the Bersimis-1 project was still under study, the young Archambault, just 30 at the time, managed to persuade his bosses at Hydro-Québec to reconsider the technology they were using for high-voltage power transmission, and switch from 120,000 to 315,000 volts.[6] He argued that this huge leap would allow Hydro-Québec to make enormous strides, reducing the number of lines being used by a factor of six. It did, however, present a real risk: there were only a few electric systems on the planet using 315-kV lines at the time, and the technology was still at the experimental stage. In order to increase the voltage of its lines, Hydro-Québec needed cutting-edge equipment including new transformers, new electric towers and its own telecommunications system that made it possible to control sophisticated equipment from a distance. The only 315-kV lines currently in use were in the United States and Europe, notably in Sweden, mostly for small parts of networks. Hydro-Québec could benefit from other utilities' experiments, but the project was still risky.[7]

Yet Jean-Jacques Archambault wasn't completely satisfied with the 315-kV solution. He thought it would be a temporary fix that would solve the challenges of the Bersimis project, but nothing after that. Behind Hydro-Québec had already begun considering the gigantic Manicouagan-Outardes project—a 6700 megawatt complex three times the size of the Bersimis complex. Studies for these new dam projects began in 1955; the Manicouagan-Outardes project was officially launched in 1959, even before the Cabinet had approved it. "La Manic," as it was nicknamed, would become the flagship project for an entire generation of Quebeckers. Yet in 1959, engineers still hadn't figured out an economical solution for transporting electricity to the southern part of Quebec.

Jean-Jacques Archambault, however, had an idea.

Years before the sound of chain saws rang out on future dam sites of the Manicouagan-Outardes project, Jean-Jacques Archambault had demonstrated that Hydro-Québec would need no less than 30 315-kV lines to transmit its electricity to the markets in southern Quebec. The vision of dozens of power lines running side by side all the way from Quebec to Montreal was harrowing, from an aesthetic and economic standpoint.

There were already a few engineers at the time arguing in favour of increasing the voltage of the lines to 450 kV; some said 535 kV, and a few others, even 630 kV. By 1958, Jean-Jacques Archambault was convinced that the only way Hydro-Québec could really reduce its transmission costs was by dramatically increasing the line voltage: to 735 kV. According to the engineer's first calculations, Hydro-Québec would only need to use two transmission lines in that case, instead of 30,[8] which would reap great savings for the utility (see Fig. 2-C). However, in 1959, when work on the Manicouagan-Outardes project was about to start, this wasn't a viable option for the simple reason that the technology didn't yet exist: there was no 735-kV lines, nor even 525-kV lines.

The clock was ticking for Hydro-Québec, whose commissioners had already set their sights beyond Manicouagan. The company was considering even larger projects in Churchill Falls, Labrador and further north in Quebec, around James Bay. These projects would be close to 1500 km from Quebec's main load centres. If transmitting electricity over long distances wasn't made practical, or profitable, to keep up with growing demand, Quebec would be forced to follow its neighbour's footsteps and build conventional

How High Voltage makes it possible to transmit more power (kV)

Voltage (kV)	Power capacity per line (MW)	Number of lines required for Manic-Outardes (6700MW)
120	40	168
230	135	50
315	300	23
735	2,100	4

Fig. 2-C: The voltage of a 735-kV line is six times more than that of a 120-kV line. But it can carry 52 times more power (megawatts, MW). It was this technical decision that allowed Hydro-Québec to develop the province's hydroelectric resources. Source: Hydro-Québec TransÉnergie.

thermic or nuclear generating stations closer to its key sectors along the St. Lawrence Valley.

The decision to develop technology to transmit electricity over very long distances would seal Hydro-Québec's fate, by allowing it to follow a completely different path than that of neighbouring grids, and avoid the stumbling blocks that had proved detrimental to these companies and their customers. But it was a difficult choice. The accepted wisdom of the time among both Canadian and American engineers was that conventional thermic and nuclear generating stations were the least expensive options. Many engineers and managers at Hydro-Québec favoured the nuclear energy route at the time. Starting in 1960, the highly reputed Shawinigan Power Company, Quebec's second-largest utility at the time (which would be nationalized in 1963) was studying a project to build a thermal generating station in Tracy, east of Montreal, a 660-megawatt monster which would run on heavy fuel oil. And there were other projects in the pipeline.

However, an internal faction at Hydro-Québec continued to push the idea that, in the long run, hydroelectricity would be Hydro-Québec's least costly option. Cost-benefit studies comparing different energy industries—the first of their kind in Quebec—demonstrated in black and white that even if the initial cost of infrastructure was higher for hydroelectricity, the operational costs would be lower since water was free. Of course, that was only true if transmission costs were also kept low.[9]

The solution Jean-Jacques Archambault was proposing would solve Hydro-Québec's distance problem. There were still technical problems to work out, but numerous leading experts from France maintained that those problems were surmountable. During a historic meeting on August 20, 1962, Hydro-Québec president Jean-Claude Lessard announced that Hydro-Québec would embark on the course of action Jean-Jacques Archambault had proposed. It was the middle of Quebec's Quiet Revolution; Expo '67 was coming to Montreal; Americans had announced they were going to the Moon, and Hydro-Québec's management wanted to turn Quebec into a trailblazer. Adventures were on the horizon.

Leaping into the Unknown

In North American electricity circles, Hydro-Québec's decision was met with shock and dismay. Quebec was the first place where a utility would carry so much voltage over long distances not just as an experiment, but for a real project. "The Americans were wondering who on earth these people were in the North who were trying such a thing!" says André Bolduc.

Ten years earlier, Hydro-Québec had been a pioneer in developing the 315-kV line, but it wasn't the only utility trying the technology out: other American and European utilities were also experimenting with it through trial and error. But Quebeckers were on their own for the 735-kV line. No other company had yet tried it; there were no standards for the lines or towers; transformers and circuit breakers that could handle such a high load had yet to be invented; and there weren't any measurement instruments available for the system. What's more, Hydro-Québec was basically a medium-sized company at the time and its 172 engineers had their hands full. "We were actually behind the times," recalls André Bolduc. "The computer systems weren't very sophisticated back then. We were still working with slide rules."

The decision to go with 735,000 volts seemed all the crazier since the Manicouagan-Outardes project had already been underway for three years by that time. Hydro-Québec was already behind. The turbines of the first station were slated to start operating in November 1965, just three and a half years away. There literally wasn't a second to waste.

Things got rolling in the fall of 1962 when Hydro-Québec invited the world's biggest electrical equipment manufacturers—at the time, General Electric, Westinghouse, ASEA and Brown & Boveri—to discuss what equipment it would need. The multinationals all had the laboratories and researchers necessary to develop the equipment Hydro-Québec would need to carry out the 735-kV project. And the companies actually reacted more positively than Hydro-Québec expected. Some still thought Jean-Jacques Archambault's project sounded crazy, but they recognized that it did hold the promise of liberating hydroelectricity from its principal constraint: distance.[10] The companies quickly mobilized their teams to take up the challenge; whoever could supply the equipment Hydro-Québec needed would, in turn, control the market for it.

Everything had to be built from the ground up. For lines to cross Quebec's rivers, Hydro-Québec would need towers as tall as skyscrapers. It would need instruments that could measure the system's activity with precision. From this scramble to develop the necessary equipment, Hydro-Québec would go on to patent a cross-arm system that could hold four wires in bundles at equal distances from one another, despite wind, snow or freezing rain.

Someone needed to come up with a system of gigantic circuit breakers that could separate two pieces of copper with enough speed and strength to prevent the high voltage current from setting off electric arcs that would damage the equipment. During the first trial of the 735-kV line in September 1965, after three years of intense work, Hydro-Québec's team of engineers spent an entire day trying to figure out how to get it all up and running. It took the team many weeks to master the process and make sure the line would work when it was inaugurated.

Two months after these trials, on November 29, 1965, Quebec Premier Jean Lesage turned the switch that put the transmission line between Manicouagan and Montreal into service. Quebec had done it. News of the innovation made so many waves in the international electric industry that Hydro-Québec's engineers ended up being invited to conferences worldwide to talk about the new line.[11] The following year, the Canadian Electricity Association awarded Hydro-Québec, collectively, the Electrical Man of the Year trophy.

However, even though Hydro-Québec had pulled off creating the 735-kV line, the technology behind it was still far from perfect. Hydro-Québec had climbed on the saddle of a bucking bull, with no idea how to control the animal. The circuit breakers were still tripping unexpectedly; no one could figure out why the voltage in the lines kept oscillating like a swing set; the lines were also more sensitive to common conditions like freezing rain, lightning and forest fire smoke, and even solar magnetism. Starting in 1967, Hydro-Québec put together a team of researchers to work on solutions to the grid's technical challenges. In 1970, it inaugurated the IREQ research centre. An entire generation of researchers kept busy for 25 years figuring out and solving the mysterious problems that troubled operations of the 735-kV line.

This real-time learning process was a roller coast ride that rattled both Hydro-Québec and its customers. All Quebeckers over 50 today recall at least a few of the system-wide outages that were

Fig. 2-D: On November 29, 1965, Premier Jean Lesage (centre) inaugurates the first 735,000-volt line. To his right, the engineer Jean-Jacques Archambault, to whom we owe the technology. Standing with glasses is Hydro-Québec's former CEO Jean-Claude Lessard, who gave Archambault permission to pursue the project.

so common in those years: during the first 16 years of the 735-kV network, there were at least a dozen, in addition to countless local outages. After a period of calm from 1982 to 1988, there were three general power outages over the next 11 months. The first, in April 1988, interrupted a hockey final between the Montreal Canadiens and the Boston Bruins. Six months later, in November, the poor Bruins were playing again in Montreal when the second outage occurred. The last major outage, caused by a solar storm, happened five months later in March 1989, and lasted nine hours (The Bruins were spared this one). This last one prompted Hydro-Québec to make some major investments in improving the whole system.

Fig. 2-E: Scientists standing like miniature figurines in front of the futuristic devices that make it possible for Hydro-Québec to master high-voltage transmission.

It was thanks to these many improvements that the system didn't collapse during the 1998 ice storm. Between January 8 and 11, 100 mm of ice formed on power lines around Montreal, bringing on the worst power outage in Canadian history, and provoking the biggest mobilization in the history of the Canadian Armed Forces. Yet Hydro-Québec's grid had become robust enough to prevent a province-wide blackout. Even 10 years earlier, such a degree of stability in the system would have been unthinkable.

The Nuclear Siren Song

The decision to transmit electricity at high voltage made it possible for Hydro-Québec to resist the siren song of nuclear energy in the 1960s and 1970s. Wholesale nuclear power was the choice of many modern societies, including Ontario, France, Japan, the United Kingdom and the United States. Hydro-Québec's CEO at the time, Jean-Claude Lessard, even became president of the Canadian Nuclear Association in 1965. The Canadian government, through the federally owned nuclear science and technology laboratory Atomic Energy of Canada Limited, pressured Quebec to build the CANDU nuclear reactor it had developed in Ontario (CANDU, short for Canada Deuterium Uranium). In 1966, the Quebec government approved the project to build an experimental nuclear generating station, Gentilly-1, in Bécancour, across the St. Lawrence River from the city of Trois-Rivières. Hydro-Québec operated the station for several years, but refused to take possession of it, and with reason: the white elephant didn't deliver on half its promises.

Looking back, one could say Quebec got it right.

Nuclear power, once the mark of a modern society, turned out instead to be a trap. Ontario Hydro thought it had been wise to build 20 reactors in three giant nuclear power stations. The reactors would turn out to be a money pit.

Starting in 1964, Ontarians decided to bank on nuclear by building a first large-scale generating station with four nuclear reactors in Pickering, a suburb of Toronto, then four more quickly, at the Bruce Nuclear Generating Station on Lake Huron. During the 1980s Ontario Hydro doubled the number of reactors in Pickering and Bruce County. Over the same decade, Ontario Hydro started building a third generating station with four reactors in Darlington, 50 km east of Toronto. This last station was a financial black hole, costing taxpayers some $14 billion, three times its original budget. By the time Darlington was in operation in 1993, Ontario had to renovate the first four reactors in Pickering and as many in Bruce County—two had to be decommissioned outright. In 1999, the Ontario government absorbed more than $20 billion of atomic debt. And the saga isn't over, since 10 of Ontario's 18 nuclear reactors need to be refurbished soon. The bill will be hefty: it will cost at least $20 billion to refurbish, or retire the reactors that are too damaged to be saved (and those will have to be replaced with a new energy source).

Quebec was right to be wary of nuclear power, but not everyone was in favour of being cautious at the time. The debate over nuclear power in the middle of the 1970s was a heated one. The Parti Québécois argued fiercely in favour of it, the most ardent advocate being future finance minister and premier Jacques Parizeau, who wrote ironically in *Le Devoir* newspaper, "Just because a river is French Canadian and Catholic doesn't mean we have to build a dam on it."[12]

Since the federal government was still pushing development of the CANDU reactor, the Quebec government agreed to build a second nuclear power station, Gentilly-2, in 1973. To be fair, oil-producing countries had launched an embargo that year that sent shockwaves across the planet—so experimenting with nuclear energy as an alternative made sense. That same year, New Brunswick agreed to build a nuclear reactor near Point Lepreau, almost identical to Quebec's Gentilly-2.

But Gentilly-2 would be the last nuclear station built in Quebec. When the Parti Québecois came to power for the first time in 1976, it had a complete change of heart about nuclear power and passed a moratorium on nuclear energy development the next year. While uranium could, indeed, produce prodigious quantities of energy, the technology had many drawbacks. The real price tag was much higher than what its advocates had promised; it turned out to be more costly than predicted to produce the heavy water that nuclear energy requires. Then there was the problem of how to store nuclear waste for centuries. Finally, the lifespan of the generators, it turned out, was much shorter than that of hydro-electric generating stations. A nuclear power plant like Gentilly-2 would only have operated for 29 years before being refurbished: if it is properly maintained, the life of a hydroelectric complex can stretch to up to 100 years.

In Quebec, the final nail in the nuclear coffin was planted in 2012. Refurbishing the nuclear generating station in New Brunswick station, Gentilly-2's twin sister, cost $4 billion, which was double the projected price. The Parti Québécois government finally decided to pull the plug on Gentilly-2, but mothballing the power station still cost Quebec taxpayers $2 billion.

Contrary to Ontario and New Brunswick, Hydro-Québec dodged a nuclear bullet by favouring the development of hydro-electricity so early in its history. Quebeckers went out on a limb and made an audacious choice to build high-voltage power lines. That made it possible for Quebec to develop a much less problematic energy source with hydro power. And hydraulic power was the basis for yet another good decision: to nationalize Hydro-Québec—not once, but many times over.

Chapter Three

The Electric Revolution

In the global electricity industry, Hydro-Québec is exceptional in almost every way. In spite of its small population, Québec is the world's fifth largest hydroelectric energy producer, behind China, Brazil the United States and Russia, but ahead of India, Norway and Japan. Hydro-Québec is by far the largest electricity producer among U.S. states and Canadian provinces: it's almost twice the size of Florida Power & Light, and second in revenues after Pacific Gas & Electric. This, despite the fact that its rates are the lowest on the continent. Hydro-Québec also ranks fifth for number of customers, after Duke Energy Carolina, Pacific Gas & Electric, South California Edison and Florida Power and Light. And finally, it is by far the "greenest" grid in North America, with almost 100 percent of its energy coming from renewable sources. Second on the list is BC Hydro, with 80 percent renewable sources.

These statistics place Hydro-Québec way ahead of the biggest electricity systems on the continent: Consolidated Edison in New York, Hydro One in Ontario and Eversource in Massachusetts. A quick look at Quebec's neighbours also shows how much Hydro-Québec stands apart with respect to energy choices and market position. In the State of New York, "only" eight players share the energy market; Massachusetts has 43 companies and Ontario, 76. It's not hard to see why these states and provinces, compared to

Hydro-Québec, have a hard time putting concerted energy policies into practice.

Hydro-Québec rose to this position through a series of important decisions made when the government nationalized electric companies between 1944 and 1963. This move gave Hydro-Québec a degree of coherency and cohesiveness unmatched by any other power grid. If it wasn't for nationalization, or the procedures that unify management of the territory, or the technology that makes it possible to build a network of 735-kV lines (see Chapter 2), Hydro-Québec would have become another public utility with thermal generating stations dependent on coal, gas or uranium. Quebeckers today would pay two or three times more for their electricity, rates comparable to those in Toronto, Boston or New York. And, of course, the largest of Quebec's power dams—those of the James Bay and Manic-Outardes complexes, and Churchill Falls—would never have been built. Nor would Hydro-Québec have its own research centre. Instead, like other suppliers, Hydro-Québec would be struggling to integrate renewable energy sources into its energy mix. If that were the case, the state-owned enterprise would certainly be in no position to pay billions of dollars in dividends to the Quebec government as it does each year. On the contrary, like other North American systems, Hydro-Québec would be grappling with the fatal combination of increasing costs and a declining customer base as users turn to self-generation. These two factors are behind the "death spiral" that is threatening many power systems.

Yet none of this is happening in Quebec, thanks to some enlightened decisions by the architects of nationalization in Quebec—or rather, "nationalizations," plural. While René Lévesque, minister of Natural Resources in the early 1960s, is generally considered as the father of electric power nationalization in Quebec, Hydro-Québec in fact had a grandfather in the person of Adélard Godbout, premier of Quebec from 1935 to 1944. Yet even at the time of World War II, nationalizing electricity was already an old idea in Québec.

The First Nationalization

The real pioneer who actually formulated the project was not a statesman, but a dentist in Quebec City named Dr. Philippe Hamel. The onset of the Depression in 1929 had sparked an anti-monopoly movement in the United States that had echoes in Quebec. In 1930 Dr. Hamel joined a municipal committee that was investigating complaints over exorbitant rates for electricity and corruption at Quebec Power, the local electricity monopoly. Dr. Hamel soon became the leading figure in Quebec's fight against the "electricity cartel."[1]

Only Quebeckers who are over 75 today will recall the chaos that constituted electric service when it all began. In the early 1930s, when the industry was a mere 50 years old, a haphazard electric grid was being built in Quebec, about the same way they were taking shape everywhere else in the world. There were dozens of companies with names like the Quebec & Lévis Electric Light Company, the Royal Electric Company and the Montreal Gas Company. In the first decades of electricity, the creation of a grid was mostly a matter of who managed to build a dam (sometimes just made out of wood) first, or who put up electric poles first (they were often just planks with wires hung on them). Companies just did whatever they could to get a foot in the market. After that, one electricity company swallowed another, until they were consolidated into powerful, arrogant monopolies. Some municipalities managed to take control of their electrical service by creating utilities like Westmount Light & Power and Sherbrooke Power & Light, but the majority were supplied by private monopolies.

In 1936, Dr. Philippe Hamel, an influential figure by then, was elected Member of Parliament for Quebec's Union nationale party. But Hamel failed to convince Quebec's new premier, Maurice Duplessis, of the merits of nationalizing Quebec's electricity. He would have to wait for the Liberal Premier Adélard Godbout to take up the torch.

An agronomist by profession, and a relentless modernizer, Godbout was one of the forerunners of Quebec's Quiet Revolution. While he was premier, he granted women the right to vote (in 1940) and made school compulsory until the age of 14 (in 1943). His decision to nationalize electricity would be among his most important legacies.

By the early 1940s, Quebec's electricity industry was controlled by a dozen private companies, along with some 50 municipal utilities and distribution cooperatives. Godbout could clearly see Quebeckers weren't getting their money's worth. Service was expensive everywhere, and companies were pocketing huge profits without reinvesting to improve their service or boost production. In Ontario, the government had already created a state-owned company in 1906; 40 percent of farms in the province had electricity by this time, twice as many as in Quebec.

Fig. 3-A: A coin electric meter operated by Southern Canada Power. The company was among those nationalized in 1963. It was still the era of prepaid electricity supply: instructions on the meter explained how to insert coins.

Godbout gradually set his sights on Montreal Light Heat & Power (MLH&P), the biggest energy company in Canada at the time, which was created in 1901 when the Montreal Gas Company merged with the Royal Electric Company. MLH&P was distributing electricity and gas to almost 50 municipalities around the city. The company was very profitable—too profitable, actually, as it was selling electricity for 5 cents per kilowatt-hour, or almost the basic rate Hydro-Québec charges 75 years later.[2] MLH&P's estimated value at the time was over $150 million, or twice the budget of the province of Quebec.[3] To maintain its outrageous profit margins, MLH&P invested as little as possible in its infrastructure and just put off building new generating stations as long as possible. That, of course, stifled economic development both in Montreal and throughout Quebec. Starting with the worst qualities of a monopoly, MLH&P proceeded to turn itself into a small kingdom, even refusing to collaborate with Quebec's provincial energy authority at the time, the Régie provinciale de l'électricité.

In the autumn of 1943, Godbout announced the creation Quebec's hydroelectric commission (the Commission hydro-électrique de Québec), soon to be known as Hydro-Québec, which would absorb MLH&P. When he wrote the founding legislation of Hydro-Québec, Godbout's legal advisor, Louis-Philippe Pigeon, modelled it on the Tennessee Valley Authority, one of President Roosevelt's emblematic projects to develop the hydroelectric potential in the American Midwest.

Godbout's bill, tabled on March 17, 1944, was ratified on April 14—after the stock markets closed. The next day, the five commissioners named to manage Hydro-Québec took possession of MLH&P's assets. The "nationalization" was really an expropriation and it led to acrimonious standoffs. In 1947, after three years of court actions and protests, the government paid MLH&P's shareholders $110 million in damages. MLH&P's customers saw the benefits of the expropriation six weeks after the takeover when Hydro-Québec announced it would lower domestic rates by 13 percent.

Fig. 3-B: On April 15, 1944, the five Commissioners of the Quebec Hydroelectricty Commission, soon renamed Hydro-Québec, took possession of MLH&P. From left to right: T.-Damien Bouchard (President), George C. McDonald, Raymond Latreille, L. Eugène Potvin and John C. McCammon.

Hydro-Québec and the Eleven Dwarves

Dr. Philippe Hamel and Adélard Godbout's dream of fully nationalizing hydroelectric power wouldn't become a reality until René Lévesque entered politics in 1960. At that time, Hydro-Québec was merely the largest of the one hundred hydroelectric utilities that were supplying Quebec's territory. These included 46 rural cooperatives, plus roughly the same number of para-municipal companies and 11 private companies.

In 1961, when he was still minister of Natural Resources in Jean Lesage's Liberal government, René Lévesque started discussing a secret project to nationalize hydroelectricity with his advisors.

His goal was to gradually fold all the electric utilities into Hydro-Québec, starting with the 11 private companies. Lévesque could see perfectly well that Quebec's hydroelectric grid was underdeveloped compared to its neighbours; the individual companies that supplied small territories were doing as little maintenance work as possible. The other problem was that there were no hydroelectric sites left to develop in the southern part of Quebec, and none of the 11 private companies had the resources to develop large projects further north. Both price and quality of service across Quebec's territory varied enormously; in Gaspésie, electricity cost 10 times as much as it did in Montreal, and in Abitibi, the quality of electricity was so bad the lights in houses blinked like old silent movies.

The Eleven

Company	Price $	Power (megawatts)
Hydro-Québec		3,802
Shawinigan Water & Power	395,400,391	1,531
Gatineau Power	124,511,600	552
Northern Quebec Power	15,955,500	90
Southern Canada Power	19,671,384	47
Quebec Power	29,012,347	28
Pouvoir du Bas-Saint-Laurent	14,321,020	11
Saguenay Electric	9,596,000	3
Électrique de La Sarre	550,000	2
Électrique de Mont-Laurier	346,500	(b)
Électrique de Ferme-Neuve	85,000	(b)
Saint Maurice Power	(a)	(a)
	609,449,742	6,066

(a) : Subsidiary of Shawinigan Power
(b) : Electricity redistributor without real generating capacity

Fig. 3-C: In 1963, Hydro-Québec was already a giant compared to other nationalized companies. Sources: Carol Jobin and Hydro-Québec.

René Lévesque knew that taking on "The Eleven," as they were called, would be difficult even though they were all much smaller than Hydro-Québec. Shawinigan Water & Power Company, for example, was as big as the other 10 companies combined (See Fig. 3-B), but still less than half the size of Hydro-Québec. Buying them out was an ambitious plan. The operation would cost $600 million, an astronomical sum at the time. And there was another complicating factor: the 11 companies were all English-speaking, dominated by Anglo-Canadian, American and British interests. Levesque knew these leaders would never sit back and watch their business get swallowed by French Canadians. And if the situation wasn't complicated enough, The Eleven were big contributors to political parties. When he heard about Lévesque's project, Premier Lesage feared he would be accused of being a socialist if he embraced it. In those years, at the height of the Cold War, being labelled a socialist could spell the end of a political career. Even the President of Hydro-Québec opposed national-ization, though for different reasons: the company had enough problems finishing the projects it had undertaken without adding another layer.

Lévesque knew what kind of controversy his project would spark. So he decided to be as straightforward as possible in his communications. He didn't just make a public announcement; explained the ins and outs of his project to representatives of Quebec's electricity industry during the opening speech at the National Electricity Week conference, on February 12, 1962. The owners, shareholders and managers of The Eleven immedi-ately retaliated, launching a series of personal attacks against Levesque. In the financial circles of Montreal's Rue Saint-Jacques (known then as "St. James Street"), the English-speaking press and Quebec's conservative Union nationale Party claimed the government were thieves and Bolsheviks.

Lévesque has expected this reaction, and was prepared. A for-mer journalist, he hit the road in order to explain to Quebeckers in fine detail the advantages of an integrated hydroelectric grid and a making a rational use of Quebec's water resources. Nationalization would bring energy prices down, he said, while ensuring more uniform and better electricity service across Quebec's territory. Nationalization would also usher in ambitious new hydroelectric dam projects. Lévesque's arguments quickly won over Quebec's intelligentsia and the general population, including the French-language press.

Fortunately, Lévesque had a powerful ally at Premier Lesage's office: Lesage's personal advisor, attorney Louis-Philippe Pigeon, who had been legal advisor to Adélard Godbout from 1940 to 1944 and was the author of Hydro-Québec's founding legislation. Though Montreal's English-speaking financial circles were resisting, Lesage was confident that the project made economic sense and that Wall Street financiers would underwrite the $600 million transaction. This greatly reduced the chances of him being accused of being a socialist.

On September 4, 1962, Lesage called a meeting of his ministers at his fishing camp in Lac à l'Épaule to debate Lévesque's project. Nothing actually came of the meeting itself. Lesage's cabinet remained deeply divided on the question of nationalization, with English speakers opposed to it and French speakers, including the premier himself, supporting it. Two weeks later Lesage called a referendum election on the question. The Liberal Party's slogan was *Maîtres chez nous*, ("masters in our own home, " a expression that refers to French speakers taking power over the hydroelectric industry out of English speakers' hands and controlling their own economy). On November 14, the Liberals won by a comfortable majority, with 58 percent of the vote.

Lesage's government then charged ahead with nationalization. On December 28, the government presented a bid to buy out The Eleven. Less than six months later, on May 1, 1963, the takeover was completed and Quebeckers were finally "masters in their own home." Some 500,000 new Hydro-Québec customers across

Fig. 3-D: A French course at Shawinigan Water & Power in 1964. After nationalization, Hydro-Québec taught French to hundreds of employees of the former private companies, including most management teams.

the province saw their hydroelectric rates fall to the same level as those in Montreal. "Hydroelectric dams now speak French," reported Quebec's largest daily newspaper, *La Presse*.[4]

Few Quebeckers know of it, but a third nationalization of electricity was carried out seven months later and involved the 80 cooperatives and municipal networks that were still left, which together supplied about 12 percent of Quebec's customers. In December 1963, Hydro-Québec bought 45 of the 46 electricity cooperatives. Over the next two decades, it gradually acquired some 30 of the remaining municipal utilities. The last remaining grid was Anticosti Island's, which only became part of Hydro-Québec in 1984.

The last phase of nationalization was actually never completed: of the original 46 electricity cooperatives in Quebec, one still remains in Saint-Jean-Baptiste-de-Rouville, supplying 16 neighbouring municipalities including Rougemont, Marieville and Mont-Saint-Hilaire. There are also nine municipal networks still in place: in Alma, Baie-Comeau, Coaticook, Joliette, Magog, Saguenay, Sherbrooke and Westmount.

Unlike The Eleven, who were forced to become part of Hydro-Québec in 1963, the Quebec government has never forced these small remaining networks to sell. This small group of survivors, together, supply 156,000 customers, or less than 4 percent of Hydro-Québec's customers (Fig. 3-E), and the law forces them to charge the same rates as Hydro-Québec.[5]

The 10 "Baby Hydros" and their customers

Systems	Customers
Sherbrooke	82,700
Saguenay	20,300
Westmount	10,200
Magog	10,000
Joliette	9,000
Coopérative Saint-Jean-Baptiste-de-Rouville	6,400
Alma	5,500
Baie-Comeau	5,000
Coaticook	4,000
Amos	2,900
Total	155,800

Fig. 3-E: The last remaining survivors after nationalization: nine para-municipal networks and one cooperative. Together, these distributors represent four percent of Quebec customers.

Electric Governance

Putting aside the question of Quebec's few remaining independent systems, the progressive process of nationalization reveals important differences between Hydro-Québec and its neighbouring grids, notably in Ontario. In many ways Ontario was far ahead of Quebec: it created Ontario Hydro in 1906, and in its early years, Ontario's model was very influential. President

Roosevelt used Ontario as a model when he decided to build the Tennessee Valley Authority, one of his New Deal projects, in 1933. Yet unlike Quebec, Ontario only partly nationalized its network: Ontario Hydro produced and transported electricity but left the distribution to individual municipal networks. The decision to partially nationalize was mostly a product of the era. In 1906, electricity was still a new industry. The transmission techniques and radio communications required to run a major distribution system didn't exist in 1906, so it simply wasn't possible to create a large centralized, unified system that produced, transmitted and distributed large quantities of electricity. By the inter-war period Ontario Hydro managed to distribute electricity to rural areas, but it never took over distribution in cities, which would be handled by up to 300 separate municipal utilities.

By the time Quebec decided to nationalize electricity in 1944, the technical landscape was radically different. There still weren't computers, but telecommunications and electronic calculation were making giant strides. All of a sudden, Ontario's semi-nationalization appeared to lack ambition. In contrast, Hydro-Québec and the Quebec government were determined to unite the province's hydroelectric resources under a single public service. From generating stations to power outlets, electrons in Quebec would travel through a single system.

It was a curious turn of events in Ontario's history of electricity: after making such formidable inroads in 1906, Ontario Hydro failed to keep pace with the technical progress and changes happening in the electricity market after that. The law that created Ontario Hydro actually had the effect of multiplying the number of distributors, transmitters and producers in the province. This made any efforts to reform the system difficult, and Ontario Hydro simply never achieved full control of its electric power market, the way Hydro-Québec did.

Ontario Hydro's financial picture started getting worse in the late 1980s. The nuclear energy adventure hurt the utility's bottom line, but the Ontario government didn't help things by refusing, to reform Ontario Hydro's governance structure , for both ideological and political reasons,. When things started to get worse in the 1990s, Ontario's very conservative Premier Mike Harris was not the type to argue in favour of more state control over electricity. So the inevitable happened: in 1999, the Ontario government split Ontario Hydro into five distinct Crown corporations, then slated the biggest two, Ontario Power Generation (responsible for generating electricity) and Hydro One (responsible for transmission and distribution), for privatization. To make this happen, the Ontario government absorbed Ontario Hydro's $20-billion nuclear debt, which its customers had already been paying off over the previous 15 years through a special 0.7¢/kWh fee. In 2015, the government then privatized Hydro One, selling 51 percent of its shares on the stock market. Today, people in Ontario pay electricity rates that are between two and three times higher than those of Quebec, and that doesn't include the billions of dollars Ontario taxpayers have had to absorb because of mismanagement of Ontario Hydro. Electricity prices got so out of control in Ontario that in 2017 the government launched a plan to reduce rates by 35 percent. Meanwhile, the government keeps using accounting tricks to push back the expiry date of this subsidy of billions.[6] And the saga is far from over: many of Ontario's nuclear generating stations have to be refurbished before 2030 at a prohibitive price. Even the way the government has decided to encourage solar energy, by offering generous subsidies, is turning out to be a drag on the finances of the government, the utility and its customers.

The case of Ontario is sadly common. In the last 30 years, most U.S. states have also chopped up their public utilities either for financial, technical or ideological reasons. (However, Ontario is the only Canadian province to have broken up its state monopoly, with the exception of Alberta, which has always functioned with private companies.) Quebec, over the same period, made regular reforms to the corporate structure of Hydro-Québec, as well as to

Fig. 3-F: René Levesque, Jean Lesage and Daniel Johnson at the inauguration of the generating stations at Manic-5 in 1968. This photo of three provincial premiers (future, former and present) is a testament to the continuity of Hydro-Québec's governance. Daniel Johnson died that night of a heart attack.

the regulatory framework within which it operates, which allowed the company to adapt to new circumstances without dismembering the network.

The first Hydro-Québec governance reform came at the end of the 1970s. When Hydro-Québec was created in 1944, it was actually a "commission" directed by five commissioners named by the government and given the mandate to independently decide on all matters of electricity in the province. The commissioners' primary task was to ensure that electricity was sold to customers at the lowest rate possible. The commission was completely independent in its decision-making, to the extent that in the 1960s, accusations started flying that Hydro-Québec was actually a "state within a state."

In the late 1970s, the Quebec government wondered if Hydro-Québec's founding law didn't limit its potential to contribute to the development of Quebec society. The finance minister at the time, Jacques Parizeau, was convinced Hydro-Québec should contribute to the province's finances and economic development, but that it should also be expected to apply Quebec's energy policies in tandem with the government.

As a result, between 1978 and 1983, the Quebec government passed a series of legislative measures that altered Hydro-Québec's mission and governing structures. The team of independent commissioners directing Hydro-Québec since 1944 was replaced with a board of directors of 11 members, each named by the Quebec government. Hydro-Québec became a company with a commercial mandate. It retained its core mission of supplying electricity to Quebeckers at the lowest cost possible, but now, like any commercial enterprise, was expected to be profitable while contributing to Quebec's economic development. In 1981, Hydro-Québec's legal status and financial structure were modified. The provincial government became Hydro-Québec's single shareholder. In the future, this shareholder would be in a position to bring Hydro-Québec's actions into line with Quebec's economic development policies, and require the company to apply Quebec's energy policies.

"It was typical of Jacques Parizeau's thinking. He had a very interventionist vision of economics," says André Bolduc, former economist at Hydro-Québec and the author of many books on hydroelectric power in Quebec. "At that stage, there was nothing surprising about creating a law to recognize Hydro-Québec as an economic tool for Quebec. It was a change that had to be made. It acknowledged the importance Hydro-Québec already had in provincial affairs. The province had to have some influence over how Hydro-Québec was making its annual investments. You could say Hydro-Québec has had two lives: one before 1981 and one after."

This "commercialization" of Hydro-Québec (employees of the state-owned company used that word in reference to the 1981 law) had consequences. Henceforth, productivity gains would not translate directly into lower rates, but become dividends, which in turn would contribute to Quebec's public finances. In 1989, a new change in the law reinforced Hydro-Québec's commercial vocation even more by giving Hydro-Québec the mandate to export electricity and carry out activities in export or energy-related fields.

Yet the years following the reform turned out to be disappointing. Hydro-Québec only generated small profits over the decade and dividends were less than expected. Hydro-Québec was coming out of a long phase of investments in large, expensive projects financed on high-interest loans. The government was also juggling competing goals. It wanted Hydro-Québec to make money, and contribute to regional development. But how much should the company decrease profits by contributing to regional development? Hydro-Québec's new vocation also met resistance from employees, not to mention discontent among the Quebec public, who were nostalgic for the glory days of a dam-building Hydro-Québec that operated in the pure spirit of public service. It took a full 10 years and a few labour strikes for Hydro-Québec's personnel to digest the changes and return the company to the kind of profitability its monopoly position should have ensured.

The next set of reforms, in the mid-1990s, was met with much less resistance, especially internally at Hydro-Québec. The round of reforms started in 1996 when the Quebec government created an energy regulatory authority, the Régie de l'énergie. The government had for some time been looking for mechanism to "depoliticize" electricity rates. Since the first days of nationalization, parliamentary committees had decided on rate increases, and the final decision came from the premier's office. In this system, decisions to increase rates could be made to meet Hydro-Québec's financial needs, but were also influenced by political factors (the temptation to please voters). With the creation of the Régie de l'énergie, rate increases became totally apolitical. The new Régie was granted quasi-judiciary powers and became responsible for regulating rates for both electricity and natural gas. Hydro-Québec, itself, had to go before the Régie's members to defend every request to increase rates. The only way government could influence the regulatory authority was by changing the laws that governed its decisions.

And there was another good reason for creating the Régie de l'énergie: it protected Hydro-Québec when the publicly owned energy supplier entered the newly deregulated energy market in the United States. At the beginning of the 1990s, the word "liberalize" was becoming fashionable in electricity circles in the United States as Americans looked to reduce energy prices by breaking up electricity monopolies. The idea was to divide up generation, transmission and distribution into distinct entities. According to this logic, any company could enter the market to produce and sell electricity. The only thing these new producers had to do was hook into a transmission and distribution system and convince someone to purchase their electricity. In theory, the electricity market would be one where supply and demand operated without constraint. However, in practice, liberalizing energy markets required the creation of new regulatory categories and a reorganization of the market into trade zones. In the end, the electricity market became more fluid, but rates didn't fall and customers didn't gain anything.

However, "liberalizing" the U.S. electricity market was good news for Quebec: it offered a way for Hydro-Québec to sell surplus energy at a significantly higher price than in the domestic market. The only obstacle Hydro-Québec had to overcome was the Washington-based Federal Energy Regulatory Commission (FERC). The FERC refused to authorize Hydro-Québec to enter the U.S. energy market if the Quebec government continued to protect its domestic monopoly. But Hydro-Québec found a way around this by changing its corporate structure to theoretically allow competition. Because electricity prices in Quebec are so low, no neighbouring utility or private generator could ever compete with Hydro-Québec on its own turf anyway. So opening the Quebec market to competition posed no risk to Hydro-Québec.

To satisfy the FERC, the Quebec government passed what amounted to a theoretical reform. In 1997, it transformed Hydro-Québec into a holding company with three distinct business units: Hydro-Québec Distribution, Hydro-Québec Transport (called TransÉnergie) and Hydro-Québec Production (The creation of the Régie de l'énergie, with its quasi-judicial powers, the previous year was another requirement of the FERC). Hydro-Québec's three business entities would function independently enough to meet U.S. market rules. The companies would be run from the same floors of Hydro-Québec's headquarters in downtown Montreal. The structure satisfied the Americans, and they opened their energy market to Hydro-Québec.

What amounted to an administrative pirouette made it possible for Hydro-Québec to cash in on a whole new export market for electricity, and has increased its export sales many-fold. It also helped electric utilities in the U.S., Ontario and New Brunswick diversify their energy sources. Hydro-Québec's regulatory tricks have also allowed the company to rake in billions more in profits and, as a result, pay record dividends to the Quebec government, year after year. All without increasing Quebeckers' already low rates.

Chapter Four

The Future Is Here

When we took our family trip to visit "La Manic" we were dumbfounded by the number of tourists who had driven 200 km into the wilderness just to see a hydroelectric dam. But there were a lot of discoveries in store for us as we drove along Route 389. On the highway linking Baie-Comeau to Labrador, 567 km straight north, many of the biggest challenges Hydro-Québec faces are in plain view.

The first of these was just 23 km north of Baie-Comeau, at the campground across the road from the Manic-2 generating station. We decided to spend the night there before driving to Manic-5. The Manic-2 station is a lot smaller than Manic-5, and not nearly as spectacular, but it still generates 1229 megawatts of power, making it one of the most powerful stations on the grid. Just as we were getting ready to settle down for the night, our curiosity was piqued by a little log cabin near the campground showers.

The cabin, with its gable roof and large windows, looked inviting, but it wasn't for rent. We later learned that the owner's son had built it for his personal use. Yet it was the series of solar panels on the cabin's roof that grabbed our attention. It struck us as a little absurd. The little cabin was literally across the road from one of the biggest, most efficient generating stations in a grid that produced the cheapest, greenest energy in the world. What exactly would motivate someone to experiment with photovoltaic solar energy here (or anywhere in Quebec at all, for that matter), especially when hydroelectricity is still cheaper than solar power, and just as green?

The small cabin at Manic-2 is a perfect illustration of one of the four biggest challenges Hydro-Québec is facing. Solar energy and other sources of energy like wind or geothermic, commonly referred to as "green," have become so attractive in popular thinking that their future growth is almost a certainty—even if they don't really make sense in Quebec. When more electricity customers do turn to self-generation (and they will), Hydro-Québec will lose sales. It sounds obvious, but it comes as a shock to a company that has never experienced so much as a dip in sales since it was founded in 1944.

Hydro-Québec has at least three other challenges on the horizon. First, the company will have to come to terms with new consumer habits that are changing the way people actually use energy. Second, the government-owned enterprise is presently at risk of running short of power—a concept difficult for customers to grasp, since Hydro-Québec actually has a *surplus* of energy. And finally, Hydro-Québec's contract with the Churchill Falls generating station in Labrador will expire in 2041: few Quebeckers are conscious of the enormous challenge this represents.

All these challenges are in plain view along Route 389.

Sun, Wind, Waves

As the little cabin in the Manic-2 campground suggests, Hydro-Québec has to learn how to accommodate "alternative energies" in the electric system, particularly solar, and probably sooner than later.

Today, at the beginning of the third millennium, new energy alternatives excite the imagination of millions of consumers. The most popular are solar and wind energy, but others are on the way, like hydrokinetic (underwater windmills), geothermal energy and tidal energy (which captures the force of water moving in tides or storms). These energy sources share the quality of being "renewable," like hydroelectricity; the difference is that unlike hydroelectricity, solar (notably) offers homeowners the possibility of becoming energy self-generators.

While rooftop solar panels have become a common sight in places like Germany, California and even Ontario, they are still rare in Quebec for the simple reason that Quebeckers have little incentive to become energy self-generators. In provinces and states where energy prices are high, photovoltaic panels are a good way for consumers to reduce their electricity bill. Governments in these places offer incentives to people who adopt solar technology to reduce greenhouse gas (GHG) emissions. In Quebec, because of hydroelectric power, the stakes are not the same: the electricity Quebeckers use is already 99.8 percent renewable—an absolutely unique situation in North America.

Yet that hasn't stopped Quebeckers from dreaming of solar power. The image of solar (and wind power) is so striking that many Quebeckers have come to believe solar energy is actually "greener" than hydroelectricity, even though all studies show the contrary.

Nor is replacing hydroelectricity with photovoltaic energy exactly the obvious "win-win" situation it's believed to be. Twenty years ago, when the United States started deregulating the energy market on the continent, Hydro-Québec had no reason to worry about new competition. Because of its low rates, Hydro-Québec cashed in on deregulation by boosting exports. However, the photovoltaic solar market will give Quebec customers the possibility of competing with Hydro-Québec on its own turf for the first time.

The risk of this happening quickly is still remote. Hydro-Québec's rates are so low that the incentive to move to solar self-generation just isn't there. The cost price of solar panels, including installation costs, still makes producing solar energy a more expensive proposition than hydroelectricity. However, Hydro-Québec anticipates that around 2023, hardware stores across the province will be able to sell solar panels at a low enough price that solar energy will be cheaper than Hydro-Québec's rates.

So the question is: when this happens, will there be 100,000, 200,000 or 1 million Quebeckers self-generating electricity, and eating into Hydro-Québec's revenues in the process? Or will Hydro-Québec actually sell and install solar panels in Quebec homes, itself? For Hydro-Québec, producing solar energy could have its advantages. It might eliminate the need to build more power lines or hydroelectric dams. Or it could allow Hydro-Québec to export its surplus energy production, or direct more of its surplus to local industries. Of course all these scenarios depend on whether solar energy turns out to be a fad. If that's the case, Hydro-Québec will be better off not getting involved in solar energy generation at all. But who knows?

The solar question has provoked fierce debate among Hydro-Québec's management, all the more because so much of the issue is about human psychology, a new and disconcerting field of activity for Hydro-Québec. In 2018, the number of energy self-generators in Quebec quadrupled, reaching 716. This is way below the 30,000 self-generators in Ontario, but the sudden surge in Quebec underlines the fact people are not interested in solar energy for strictly economic reasons. Hydro-Québec will have to find a way to accommodate the "do it yourself" spirit driving the solar energy market. The company will have to accept the fact that it can't predict how many customers will be tempted by self-generation in the short, medium or long term. A single major power outage or ill-advised rate hike could prompt many of Hydro-Québec's customers to install solar panels.

In the energy business, passing fads can have a serious impact. Hydro-Québec's advertisements from the 1950s and 1960s show how hydroelectricity was associated with "liberating" women from housework at the time. In 1965, Hydro-Québec's in-house magazine

Fig. 4-A: "My best friends: the automatic washing machine, the electric drier." In the 1950s and 1960s, private and public electricity companies, including Hydro-Québec, marketed electricity as a tool of emancipation for women.

Entre Nous had ads for special courses that the company offered to teach women how to use electricity. In the next decade, the oil crisis encouraged a massive shift to electric heating (both water and air) in Quebec homes. Hydroelectricity was touted as inexpensive, clean and easy to install and manage. Yet in rural areas, where firewood was still cheap and abundant, electrification still didn't make sense from a strictly financial perspective. But people adopted it anyway. In other words, if hydroelectricity managed to win out over less expensive alternatives 40 or 50 years ago, who is to say that half a million consumers won't turn to self-generation in the next decade?

Anticipating such changes in consumer mentalities, Hydro-Québec has launched large research programs that combine technology and sociology to try to predict which model of energy consumption Quebeckers will choose, and what impact this will have on Hydro-Québec's system. The fact is, even if photovoltaic solar energy fails to take off, another facet of the energy transition has already taken root in Quebec—new consumer habits are reducing energy demand.

Stagnating Demand

On the way to Manic-5 on Route 389, at exactly kilometre 94, the spruce forest gives way to an endless landscape of electric towers, poles, wires and circuit breakers.

This is Hydro-Québec's Micoua substation, supplied by nine generating stations along the Manicouagan, Toulnustouc and Outardes rivers to produce a grand total of 8300 megawatts of power. Electricity arrives here in 120-kV and 315-kV lines. Micoua's transformers then boost the voltage to 735 kV, and send it back into the grid so it can make its way to Quebec City and Montreal. The Micoua substation will soon have a brand new high-voltage line carrying electricity to the Saguenay-Lac-Saint-Jean region. Hydro-Québec is building this line specifically to "free" energy from Quebec's North Shore, where energy production has increased over the last 20 years because of new generating stations on the Toulnustouc, Sainte-Marguerite and La Romaine rivers. Why does energy need to be "freed"? Because energy consumption in the region barely increased over the same period. That means the existing lines in the area are saturated and Hydro-Québec has had to look for a way to send the energy elsewhere.

Stagnating demand is the second-biggest problem Hydro-Québec is facing at the moment. Unlike the issue of competing with new energy sources, which is still somewhat hypothetical, stagnating demand is already having a profound effect on the company.

Hydro-Québec has so much surplus energy at the moment its reservoirs are filled to the brim. It has even been forced in some cases to open floodgates and discharge water, something the company has almost never had to do in the last 50 years. In 2018, for example, Hydro-Québec opened spillways and let billions of cubic metres of water flow out, enough to produce 10 terawatt-hours of electricity. It literally poured the equivalent of 5 percent of its annual production down the drain.

The media tends to lay the blame for this waste on new wind power contracts that force Hydro-Québec to purchase energy from private producers (while the company discharges water from its dams). In fact, the opposite is true: changing consumer behaviour is the reason for both the wind power contracts and the energy surpluses.

Since 2007, energy consumption in Quebec has not changed *at all* (Fig. 4-B). It's a totally unexpected situation that throws into question all of Hydro-Québec's predictions—predictions that were behind the decision to build windmills and new dams in the first place. To some extent, stagnating demand can be explained by the difficulties of Quebec's pulp and paper industry, which has slowed down more than predicted, and by the aluminum industry, which hasn't grown as much as expected due principally to increased competition from China. Massive deindustrialization in Quebec is another factor, as is the province's aging population. But whatever the reasons, demand for electricity is simply not increasing.

Sales at a standstill

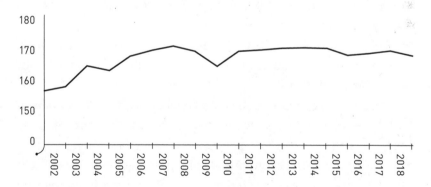

Fig. 4-B: A global phenomenon: since 2007, demand for electricity in Quebec has stagnated while the economy has grown almost 15 percent, the result of gains in energy efficiency and deindustrialization, among other factors.

Hydro-Québec isn't the only utility having problems predicting energy demand. Electricity companies all over the planet are grappling with a problem known as "decoupling": economic growth no longer goes hand in hand with a growth in energy use. The two have been paired since the appearance of the first electricity grids in the 1880s. There have been variations from year to year, from one country to another, but generally speaking, if an economy grew 2, 3 or 5 percent, demand for electricity grew at the same rate. Beyond unforeseen economic circumstances, economic growth was the primary factor used to predict growth in energy requirements. Yet that hasn't been the case anymore since 2008. Energy demand has remained stable. In 22 of the European Union's 28 countries, energy demand has actually decreased. "In the United States, this energy decoupling started as early as 1998. In Ontario, it started in 2005. And Quebec, beginning in 2007," explains Alexandre Deslauriers, Team Leader, Load Forecast at Hydro-Québec. Since 2007, Quebec's economy has grown 15 percent. "But Quebeckers' yearly energy consumption has remained stable since 2008."

Between the nationalization of electricity in 1962 and 2007, Quebec's energy consumption increased 20-fold from 8 to 170 terawatt-hours, or 7 percent per year over 45 years. This was mostly a result of economic growth, but the fact that Quebec's electricity market was underdeveloped prior to nationalization was also a factor, in addition to the mass switch to electricity for heating and industry.

In Quebec, the main "culprits" of energy decoupling have been better energy-efficient windows, insulation and lighting. Slowing population growth has also played a role as well as deindustrialization common to the economies of all developed countries, which reduces energy demand. Global warming, which is happening more quickly than predicted, is also lowering energy demand for heating—a situation not without its irony. The relative importance of these factors varies considerably from one country or region to another, but the overall impact is the same across developed nations: demand has levelled off, and things won't be any different in Quebec.

Forecasters are using technology to its fullest to try to discern new trends in energy demand. They are attempting to identify the new factors that influence changes in energy demand so they can rebuild the models they use to make predictions. The questions are wide-ranging. Will photovoltaic solar panels really spread in Quebec? Will new batteries being developed enable industry or residential customers, or both, to store electricity? What will Hydro-Québec learn from experiments with micro-systems in Lac-Mégantic, or from the experiment with solar panels in Quaqtaq, Nunavik? (See Chapter 6.) Will electric cars spread as quickly as predicted, or not? And what about home automation? Will the cable distributors of the future offer homeowners the possibility of saving money with systems that manage heat and lighting from a distance? What, for that matter, will normal climactic conditions even look like, as global warming progresses?

Since Quebeckers' strong appetite for electricity has ensured Hydro-Québec's growth and profitability until now, electricity decoupling definitely poses a threat to the future of the government-owned company. Hydro-Québec has had a commercial purpose for the last 40 years. Simply put, in a context of stagnating demand, it has to find new ways to make money. Inflation and new consumer expectations (including everything from underground power lines to better customer service) are increasing Hydro's costs. And there are limits to how much Hydro-Québec can cut expenses by being more efficient: Hydro-Québec already sells twice as much electricity as it did 30 years ago with the same number of employees.

With the trend of stagnating demand likely to last, Hydro-Québec is working to develop new markets inside Quebec (electrified transportation, greenhouses, and data centres) and for export. The company is hoping Quebec's aluminum smelters, which account for a full 5 percent of overall electricity demand in the province, will renew their supply contracts when current contracts expire in the 2030s and 2040s. However, the United States has its own problem of energy surpluses, the result of an explosion in shale gas production in recent decades, which have made the country almost self-sufficient in energy. This surplus of natural gas has naturally driven down energy prices in the American market, which means Hydro-Québec has a harder time finding buyers for its energy at a good price. In its discussions with other utilities, Hydro-Québec can still cash in on the flexibility of hydroelectric power, but the competition is fiercer than it was 15 years ago.

The third problem Hydro-Québec is facing may sound contradictory even if it's not. While it has an energy surplus, Hydro-Québec is simultaneously grappling with a shortfall in *power*.

The Power Shortfall

When we finally arrived at Manic-5, we were struck by the majestic, raw beauty of the dam's soaring cement arches. Manic-5's Daniel-Johnson Dam holds back 2000 square kilometres of water. The reservoir is so enormous that it took the Manicouagan River 13 years to fill up. The dam is unusual in that it feeds not one, but two generating stations built on opposite sides of the river. The first, opened in 1970, produces 1600 megawatts of power. The second, built 20 years later on the west side, added another 1000 megawatts for a total of 2600 megawatts of output. The name of the second station is Manic-5 PA. The "PA" stands for *puissance additionnelle* (added capacity). That perfectly expresses the station's function. In engineering jargon, Manic-5-PA is called "over-equipping." It's added capacity designed to boost the total power the complex can supply in order to meet increasing demand.

Yet Manic-5 perfectly illustrates the problem of the power shortfall Hydro-Québec is facing.

Few people fully grasp the difference between power and energy, but the distinction is at the heart of any power system's job.

"Power" refers to the intensity of the energy that is provided. It's the load at any moment, calculated in watts (or kilowatts, megawatts or gigawatts). Energy, on the other hand, is the amount of power used over a given period of time, expressed in watt-hours (or kilowatt-hours, megawatt-hours or terawatt-hours).

One may wonder, fairly, how Hydro-Québec can possibly lack power when, as we've seen, it has a surplus of energy. Take the example of the two generating stations at Manic-5. Together, they produce 2600 megawatts of power, enough to supply Quebec's nine aluminum smelters (which, combined, actually do require 2600 megawatts to operate). The quantity of energy stored in the reservoirs that feed the generating stations, calculated in cubic metres, is one of Hydro-Québec's commercial secrets, but we know there is enough to keep nine aluminum plants running for many years to come.

But if Quebec were to build a tenth aluminum plant, the two generating stations at Manic-5 would not be able to supply enough energy to it, even though there is extra water in the reservoir. That's because the ability of the generating stations to turn that energy into power, measured in watts, is set. To make more power, Hydro-Québec would have to build another generating station. Yet doing that means Hydro-Québec would run the risk of emptying its reservoirs too quickly. That might put Hydro-Québec in a situation where it is generating too much power without the necessary water storage (energy) to draw on. The challenge for any electric utility is finding the right balance between having energy available (in Hydro-Québec's case, in reservoirs) and having the means necessary to transform it into power (in generating stations) and deliver it to customers.[1]

Hydro-Québec's system has 63 generating stations in all. With an additional 1900 windmills, as well as electricity it purchases from neighbouring utilities, Hydro-Québec can theoretically generate over 47,000 megawatts of power. In reality, the number is closer to 40,000 simply because not all generating facilities are available at the same time. There is always a turbine, or a line or a dam somewhere in the system that is out-of-service, whether it's under maintenance, or because there isn't enough wind to turn the windmills, or the water level of a river is too low. Hydro-Québec also has to have extra capacity available for any type of emergency situation that might arise.

While 40,000 megawatts seems like a lot of power, it's just enough to meet the needs in Quebec during peak demand period, which is in the winter when everyone is running heaters and turning on lights, stoves, dishwashers and water headers roughly at the same time—and that's not counting extras like raclette

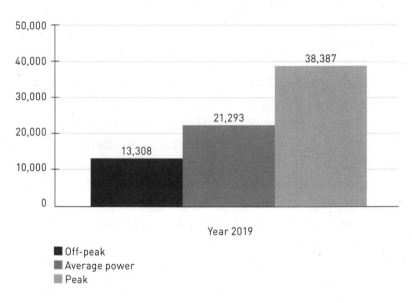

Large power fluctuations

Year 2019

- Off-peak
- Average power
- Peak

Fig. 4-C Hydro-Québec has to deal with enormous fluctuations in demand. Average yearly power demand is 21,300 megawatts. In the summer, off-peak demand can be as low as 13,300 megawatts. In periods of extreme cold, when everyone is heating, peak power demand (38,400) is almost double the average
Source : Hydro-Québec.

grills, Jacuzzis, Christmas decorations, computers and cell phones. Add to that the famous "ghost charge" of electric appliances we believe are off but that draw electricity anyway, and household consumption jumps another 10 percent.

All electric power systems have to contend with this "problem" of having enough power because their capacity to meet energy needs is limited. If any entire population, anywhere, turned all its appliances on at the same time, no power system would be able to supply enough energy to make them all work. The result would be a power outage. In the case of Hydro-Québec, this power challenge is amplified by the fact that since the 1970s houses have been using electricity for heating and hot water, which force the utility to generate even more power during peak periods (when the temperature dips, everyone requires more power for heating at the same time). Hydro-Québec calls these periods "the critical hours" and they vary from 100 to 300 hours over the winter, depending on how cold it gets. Hydro-Québec doesn't have much leeway to meet exceptionally high demand. Variations in demand, for that matter, can happen very suddenly. On December 28, 2017, for example, a Siberian cold snap hit Quebec in the middle of the Christmas holidays. At exactly 4:58 p.m. Hydro-Québec was supplying 38,420 megawatts of power. Two minutes later, demand had fallen to 38,204 megawatts. The sudden variation of nearly 220 megawatts is enough power to light up a city the size of Sherbrooke (population of 160,000).

Better construction standards and increased energy-efficiency measures do mean people spend less, overall, on heating today. At the same time, we are plugging more and more devices into the system, whether it's wine cellars, iPhone chargers, pool heating systems, chalet furnaces or electric cars—many of which people also tend to turn on at the same time of day. When they are added up, these new appliances account for 100, 200 or even 400 extra megawatts of power per year.

Since 1944, Hydro-Québec, like most other large power system operators, has solved the power challenge by barging ahead more or less blindly. In Hydro-Québec's case, this has meant building more dams. When demand rose above a certain ceiling, the company quickly built another dam and added new generating stations, high-voltage transmission lines and transformers. It was a bit like trying to solve traffic problems by adding extra lanes to existing highways, overpasses, bridges, streets and even country roads.

This old-school solution—keeping up with demand by building more, bigger infrastructure—has reached its limit today. For one, it's expensive. Hydro-Québec has had to add a third transformer to each substation throughout the province to ensure it can supply enough electricity to homes and businesses during peak hours. The third transformer costs the same price as the first two, but is only used a few hours per year. The same logic applies to any more equipment Hydro-Québec has to add. To go back to the highway analogy, when the fifteenth new highway lane is only being used 1, 2 or 3 percent of the time, it's time to rethink how transport infrastructure is organized. The same logic applies to electricity transmission.

Hydro-Québec is also facing two relatively new challenges: the increasing requirements of "social acceptability" of its projects, and the new phenomenon of varying generation.

Legitimate as they are, social acceptability requirements put a brake on Hydro-Québec's ability to meet higher power requirements. Two generations ago, few in Quebec ever questioned the decisions the government-owned utility made about building new dams or power stations. That's all changed. To build a new generating station or power line, or even to increase its rates, Hydro-Québec has to first make sure its plans comply with social and environmental acceptability requirements. In a famous case that started in 2013, the mayor of the town of Saint-Adolphe d'Howard, in Quebec's Laurentians region, led a movement to force Hydro-Québec to either bury a new high-voltage line that was slated to pass through the village, or cancel the project altogether. The town didn't succeed in stopping the project (and even if it had, it's doubtful the Régie de l'énergie would authorize Hydro-Québec to increase its rates to pay for the extra costs of burying a high-voltage line). In other words, to some extent Hydro-Québec has its hands tied: the population is mobilizing to protest new projects at the very moment the company faces power shortfalls.

Adding to that, Hydro-Québec has to deal with the new challenge of varying generation. Once upon a time, Hydro-Québec had a single objective: supplying enough power to keep up with demand in the province. For decades, Quebec's hydroelectric grid had an almost miraculous ability to accomplish this on short notice. It takes only eight minutes to get a turbine operating, and five minutes to shut it down. (By comparison, it takes hours to ramp up a steam-operated generating plant.) But the addition of alternative energy sources (for example, windmills and solar panels) to the grid has ushered Hydro-Québec into a new era where generation, like demand, now depends on the weather. Hydro-Québec presently gets 3600 megawatts of power from windmills. That's 10 percent of its overall production, a huge proportion—on paper. Yet in practice, Hydro-Québec considers itself lucky when wind conditions allow it to generate a third of this power. There just isn't enough wind all the time. That means Hydro-Québec has to have some dams and turbines available for the single purpose of ensuring supply consistency. The problem will get worse if solar energy spreads in Quebec. This is the paradox of energy sources, like wind and solar, which don't have reservoirs: Hydro-Québec not only faces the challenge of meeting varying demand; it now has to contend with varying generation, which eats into its capacity to meet increasing power requirements.

The new unknown of varying generation capacity, which actually increases Hydro-Québec's power deficit, makes the search for new solutions more urgent than ever. Simply put, Hydro-Québec is hitting its generation ceiling. "We have enough power at the moment, but the deficit increases from year to year and is projected to reach 1150 megawatts by 2024," says Alexandre

Deslauriers. A sign of the times: in 2017, Hydro-Québec started up its gas-fired plant in Bécancour for the first time. The enormous kettle of 411 megawatts had never been used before for the simple reason that Hydro-Québec had enough power reserves until then. That changed on December 22 and 28, 2017, when the peak in demand was so high that Hydro-Québec had to run its steam-operated generator for a mere hour to make up for a 100-megawatt shortfall. If Hydro-Québec doesn't find another way to solve its power shortfalls, these kinds of situations will happen more frequently and will be more intense.[2]

Hydro-Québec is considering various solutions, including putting new fee structures in place with time-of-use rates, and new practices to encourage consumers to use power outside of peak periods, such as cutting power to water heaters or controlling home thermostats from a distance. Other possible solutions include encouraging customers to store their own energy using batteries.

When it comes to the problem of power shortfalls, Hydro-Québec doesn't have much choice but to act. One more factor makes the situation even more pressing: the looming end of Hydro-Québec's contract with Newfoundland over the Churchill Falls generating station in Labrador.

The Big Importer

When we visited the Manicouagan River area, we didn't drive all the way to the end of Route 389. That would have taken us to Newfoundland, where the road becomes the Trans-Labrador Highway. Had we driven along that highway to kilometre 261, we would have reached the Churchill Falls generating station, from which Hydro-Québec has been buying almost all the energy produced since 1976.

In the collective imagination of Quebeckers, Hydro-Québec is a great energy exporter. In reality, it only exports in good years.

The rest of the time, it is an important *importer* that buys 31 tera-watt-hours of electricity from Newfoundland. That's 20 percent of Quebec's overall energy consumption.

Churchill Falls is an exceptional site. A series of waterfalls and cascades that fall 312 metres over a 25-km distance at the rate of 1400 cubic metres per second, its generating station produces a total of 5400 megawatts of power. That's double the amount of power Manic-5 produces with the same water flow. Yet the most astonishing thing about Churchill Falls is the price tag: Hydro-Québec pays Newfoundland the very modest sum of 0.2 ¢ per kilowatt-hour (two tenths of a cent, called "mills" in engineer language), a very low rate that corresponds to one fourteenth (7 percent) of the average cost of generation for Hydro-Québec. Every year, Hydro-Québec Production buys $62 million of electricity from Newfoundland, then resells it to Hydro-Québec Distribution for $900 million, and even more on the export market.

Unfortunately, all good things come to an end (though, of course, Newfoundland doesn't see it this way). Hydro-Québec's

Fig. 4-D: Churchill Falls. In 1966, before the dam was built, and in 2008, after the water had been put "to work" to produce electricity.

contract with Churchill Falls will expire on August 31, 2041, in 22 years. That's not a lot of time in the electricity infrastructure business. To understand the choices Hydro-Québec will have to make, it's best to start with the story of how this extremely lucrative arrangement for Quebec came about.

How did Hydro-Québec get involved with the Churchill Falls project in the first place? Simply put, the project would not have seen the light of day without Quebec's help. Newfoundland started thinking about exploiting the site's potential in 1952. The problem was, the hypothetical capacity of a generating station on the Churchill Falls site largely exceeded the energy needs of the province. Newfoundland knew it would have to sell the energy to Quebec, the United States or to Ontario to make it worth the cost of building a dam. The problem was, in 1952, the technology to transmit so much energy over very long distances at an affordable price did not yet exist.

In 1965, Hydro-Québec changed the equation by putting its new high-voltage transmission technology into service in the form of the 735-kV line (see Chapter 2), which made it possible to transmit large quantities of energy over long distances at a low cost. In 1966, Hydro-Québec came forward with a proposal to not only buy energy from Churchill Falls, but to finance the construction of the dam and pay for the transmission lines out of its own pocket—a total cost of $1 billion (in 1969 dollars, the year the contract was signed). According to the contract, starting on September 1, 1976, Hydro-Québec would buy 95 percent of the generating station's output for the next 40 years at an average price of 0.25¢ per kilowatt-hour. In 2016, Hydro-Québec applied its option to renew the contract for another 25 years at 0.2¢ per kilowatt-hour.

The contract became a political hot potato in Newfoundland the moment it was signed. Newfoundlanders felt Quebec had taken them to the cleaners. The rate, non-indexed, corresponded to market prices for energy in 1969, when the reference price on the U.S. market was 0.3¢ per kilowatt-hour. That was before the high inflation of the 1970s and 1980s. Newfoundland challenged the agreement, dragging Hydro-Québec to court 17 times in an attempt to break the contract. Three times, in 1984, 1988 and 2018, the Supreme Court of Canada returned the same verdict: the contract was valid; it had been negotiated in good faith, and it was rock solid.

Hydro-Québec has just started talking openly about what will happen when the contract with Churchill Falls expires in 2041. It will be a shock to the company, but to what degree will depend on the good will of Newfoundlanders.

There are three stakes: technical, financial and rate related. If Newfoundland decides to stop selling its electricity to Quebec, Hydro-Québec will immediately be 5000 megawatts short of power and 31 terawatt-hours short of energy, roughly 20 percent of Quebec's overall energy consumption and a total of 12 percent of its overall energy requirements. From a technical standpoint, the disappearance of this source will be like an electric shock treatment for Hydro-Québec. The company has enough difficulty meeting current demand as it is; in this case, there will be almost nothing left over to export.

However, this scenario—a complete retreat by Newfoundland—is not very likely for the simple reason that Newfoundland has neither the ability to consume all the energy it produces, nor the means to transmit the electricity to other provinces or U.S. states without going through Quebec. The only thing that would make a full retreat possible would be for Newfoundland to open extremely high-energy consuming factories on its own territory. In either case, Quebec will have time to see what's coming. Since 1963, Hydro-Québec has owned 34 percent of the Churchill Falls (Labrador) Corporation (CFLCo), with the rest belonging to Newfoundland.[3] But Quebec has to be prepared for the possibility of losing even part of the electricity it buys from Newfoundland. "The good news is that we're talking," says Éric Martel, President and CEO of Hydro-Québec. "We have always kept the dialogue open, even during the lawsuits, but now that Newfoundland has exhausted all its possibilities of legal action against Quebec, our discussions will be more productive."

The most likely outcome is that the contract with Newfoundland will be renewed, in part or totally, at a new, considerably higher price. This will be a financial shock for Quebec. In 1998, in discussion over the Gull Island project in Labrador downstream from Churchill Falls, Quebec Premier Lucien Bouchard and Newfoundland Premier Brian Tobin actually buried the hatchet by announcing a new joint project that would have been a model for integrating two systems, but the project never materialized. However, the two governments did agree in the process to update the price of energy from Churchill Falls to 4.7¢ per kilowatt-hour, or 23 times the 1969 price.[4] Quebec knows that if Newfoundland continues to export its electricity after 2041, the future price will very likely be even higher. Negotiations promise to be long and difficult. "I'm not expecting any gifts from Newfoundland," François Legault told us in an interview.[5]

Quebec customers will likely absorb the difference, and the shock of higher rates will hit them hard. So the end of the Churchill Falls contract creates a dilemma: Quebec either loses access to Churchill Falls or has to buy electricity for 23, 30 or maybe 35 times the price it was paying. Hydro-Québec's profitability will decrease by half, which means smaller dividends for its only shareholder—the Quebec government—and significant shortfalls in the province's finances. That is, of course, unless the Quebec government decides to just pass the bill on to Hydro-Québec customers. Although Hydro-Québec's 34 percent ownership in Brinco means Hydro-Québec is entitled to 43 percent of the proceeds of the much higher electricity prices. The rate shock in that case could translate into a 15 to 20 percent increase for customers. There have been public outcries in Quebec over smaller hikes than that.

But there is another option: Hydro-Québec could just take a pass on renewing the Churchill Falls contract and instead, explore other avenues like new energy efficiency measures and large-scale controls over power. Or it could consider launching new projects, whether hydroelectric dams, or solar or wind parks. Éric Martel also believes it's possible to start talking to Newfoundland about joining forces to take advantage of opportunities in the U.S. energy market. It's just one of many options Hydro-Québec is studying.

Chapter Five

Do-It-Yourself Power

When we first set eyes on Yves Poissant's house a short distance from the slopes of Mont-Saint-Bruno, east of Montreal, we thought it looked pretty much like every other house on the street. It was only on closer inspection of the home—a spacious, two-storey suburban family home with brick facades, beige vinyl siding, a trampoline and an above-ground pool in the back yard—that we saw what set it apart from neighbouring homes. Actually, we just walked into the backyard and saw that a third of the south side of the roof was covered with solar panels.

Among energy consumers, there's certainly nothing ordinary about Yves Poissant. He has a PhD in Physics and works for CanmetENERGY, a laboratory of National Resources Canada in Varennes, where he is part of a group that works on bringing renewable energy to communities in Canada's Far North. Poissant is a specialist in photovoltaic panels, and when he built his house in Saint-Basile-le-Grand in 2008, he wanted to put this knowledge into practice.

Poissant describes his house as "an experiment." Standing in the backyard, he started by explaining the difference between the two types of solar panels we could see on his roof. On the right side are 10 photovoltaic panels that transform solar energy into electricity, which is then sent to the electric panel in his house. On the left are two large thermal solar panels: the sun heats a bundle of tubes in the panels filled with water and antifreeze, then the heated liquid runs down to the basement of the house where it envelopes Poissant's water heater, reducing the amount of electricity required to heat his water.

Yves Poissant was one of only 716 energy self-generators in Quebec at the time of writing. He is conscious he's a something of a pioneer. In the history of electricity, photovoltaic solar energy is still a relatively new field. Thanks to technology developments solar energy is now making it technically possible for energy consumers to generate their own electricity in large quantities for relatively little effort and cost.

Hydro-Québec is watching this development closely. In 2017, there were no more than 150 self-generators plugged into Quebec's network, most of them using solar panels. A few outliers had installed wind, geothermal or biomass energy producing systems in their homes. Yet in 2018, the number of self-generators in Quebec suddenly quadrupled, to 716. It's still a marginal phenomenon that represents only one Hydro-Québec customer in 6000. And it's only a fraction of Ontario's 30,000 self-generators. However, their sudden multiplication in Quebec is a sure sign that something is going on behind the electric meters across the province. The question is: what?

It's clear that the number of self-generators is suddenly growing, but their projects have also gotten more ambitious, to the point that self-generation is no longer limited to home owners. In 2009, Concordia University's John Molson School of Business inaugurated the first "solar wall" with a surface of 300 square metres that heats air and produces electricity. Concordia remained a unique case for many years. Then in 2017, in Candiac, on Montreal's South Shore, a real estate promoter announced a pilot project for a development where half the 148 townhouses would be fitted with six solar panels each, enough to produce 2000 kilowatt-hours of energy per house per year. That would be enough to lower the electricity bill of each unit by 5 to 10 percent, i.e., $120 a year. Then in March of the same year, the Quebec department store chain Simons opened the first "energy self-sufficient" store in Quebec City equipped with 3200 solar panels and a geothermic energy generation system. Simons' surplus energy will feed the shopping centre where it is housed, Les Galeries de la Capitale— another first.

A few streets away from our house in the east Montreal neighbourhood of Rosemont, there is a project called Celsius, in which 70 families on three streets share a communal geothermic heat exchanger that replaces gas and oil-powered furnaces. In January 2019, the City of Montreal announced it would be building a "net zero" welcome pavilion (meaning the total amount of energy used by the building on an annual basis is roughly equal to the amount of renewable energy created on the site) in a park in the city's Ahuntsic-Cartierville borough. In short, something has definitely been set in motion.

Fig. 5-A: Two experimental houses of the Laboratoire des technologies de l'électricité, another campus of IREQ, in Shawinigan. The introduction of solar generation in homes throws into question the very way the power grid is organized.

Even if these construction projects are quite marginal in the grand scheme of things, Hydro-Québec is expecting the alternative-energy trend to grow—especially for photovoltaic energy, since the price of solar panels is forecasted to fall. According to data compiled by CanmetENERGY, between 2001 and 2011 the price of solar panels dropped 84 percent, and has fallen another 11 percent per year since then. An internal Hydro-Québec report predicts that by 2023, the cost price of residential solar energy will fall below its basic electricity rate of 5.91¢ per kilowatt-hour.

Popular interest in solar energy certainly seems to be growing. Yet while we were researching this book, we were a bit stunned to hear about the heated debates solar energy sparks among Hydro-Québec's employees. It's an understatement to say that most Hydro-Québec employees are mystified by Quebeckers' interest in solar energy. After all, Hydro-Québec has more than enough renewable energy—energy that is still cheaper than solar by a long shot—to meet Quebeckers' current needs. Solar panels are really just repeating what Hydro-Québec already does on a larger scale, in a more reliable way, for a better price. Yves Poissant himself is a bit puzzled by the popularity of solar energy. "I was motivated more by curiosity than environmental convictions when I built this house. The fact is, electricity in Quebec already comes from a completely renewable source."

For the time being, even if the energy from the sun is free, the process of converting it into usable electric power is quite expensive. Solar panels are pricey, and so is the cost of installing, maintaining and repairing them. "This project wasn't about saving money," confesses Yves Poissant, who invested $30,000 in the system, which is 10 percent of the overall construction budget of the house. "I really did it for fun, and for the challenge."

Yet the fact that solar energy offers neither economic nor ecological advantages today (in Quebec) matters little to those who embrace it. "My house is a living thing. The sun comes in through the windows, the heat pump starts and recharges my car batteries in the garage," Poissant told us. (He has been driving a Chevrolet Volt since 2012.) Poissant then pulled his smartphone out of his pocket to show us the application he uses to calculate his solar energy generation in real time. The sky was cloudy the day we visited so the panels were only producing 0.278 kW of electricity (or 278 watts), enough to light the LED bulbs in his house. But when the sun shines and conditions are right, Poissant's house can produce up to 1300 watts, enough to power a toaster, or recharge his car.

However, even if neither the ecological nor economic motives behind solar energy are clear at the moment, nothing changes the fact that energy self-generation in Quebec is growing. For the first time in its history, Hydro-Québec is facing the possibility of losing its monopoly on energy supply some day. As Hydro-Québec President and CEO Éric Martel explained to a group of international leaders and businesspeople at the International Economic Forum of the Americas held in Montreal in June 2017, "Self-generation in solar energy is a trend all over the planet and we can't ignore it. Certain customers will want to self-generate out of principle, because it reflects their values."

When we met him a year and a half later, Éric Martel had obviously thought more about the issue and was more nuanced in his remarks. "I think there will be solar panels in Quebec, but the movement will build much slower here than elsewhere for the simple reason that consumers don't feel the same sense or urgency as they do in California, where electricity costs are very high. Electricity is still so inexpensive in Quebec that it takes more than the lifespan of a solar panel to break even on the investment in solar energy."

Quebeckers also now have access to an array of methods to reduce their energy consumption, from batteries to advanced home automation technology, geothermal energy and simple energy-saving measures. These new tools are making it possible for customers to be more autonomous and gain more control over their energy consumption, as well as save money. Together, they add up to a small revolution that is radically changing Hydro-Québec's traditional role of making sure there are enough wires and poles to reach its captive clientele. Instead, Hydro-Québec is now in the era of the "participating" customer.

Whatever technology "participating" customers decide to use, their numbers are projected to reach the hundreds of thousands by 2030. If energy-saving initiatives end up reaping savings of 10, 12 or 15 percent of overall consumption, Hydro-Québec will lose that much in sales. And the dividends it pays to the Quebec government will decline at the same rate. As a result, the company has to be prepared for what's coming.

"I don't understand why Hydro-Québec is encouraging energy self-generation," says Normand Mousseau, Physics professor at the Université de Montréal and co-chair of the Quebec Commission on Energy Issues in 2013. The researcher finds Quebeckers' interest in alternative energy sources a little exasperating. "There's no good reason to encourage solar energy generation in Quebec. The economic and environmental advantages just aren't convincing."

"These technologies could either cannibalize, or help Hydro-Québec," says Vincent-Michel Duval, Engineer in charge of integrating new technologies at Hydro-Québec Distribution, "On one hand, alternative energies could end up decreasing our revenues, while on the other, they could potentially help us solve our problem of power shortfalls and generate energy surplus, which would make it possible for us to push back certain investments in infrastructure." But for that to happen, Hydro-Québec has to propose programs that will work for both consumers and for Hydro-Québec. And to do that, the company has to understand not only the technologies being used to reduce energy consumption, but how self-generators and participating customers think.

Getting inside customers' heads is a real "first" for Hydro-Québec. Since it began in the 1880s the electric power industry has been dominated by engineers, professionals who prided themselves on making decisions that combined technical expertise with sound economic judgment. It was the engineer who decided Hydro-Québec's future by figuring out what was possible and what was profitable. Now, changes in consumer behaviour are bringing into the equation a system of values that doesn't necessarily square with conventional "common sense," whether financial or technical. So for the first time in its history, Hydro-Québec has to follow the trends set by its customers and adapt itself to their behaviour and aspirations, instead of the other way around.

To do that, the engineers at Hydro-Québec have to understand that consumers don't think like engineers. "People didn't adopt the electronic thermostat because it would cut their electricity bill by 10 percent. In some cases that wasn't even a consideration," explains Jocelyn Millette, Researcher at Hydro-Québec's Energy Technology Laboratory (LTE). This division of IREQ, based in Shawinigan, has the double mandate of studying new technologies and consumer mentalities. "Hydro-Québec succeeded in selling millions of electronic thermostats because they were attractive and also because they helped couples avoid arguments about what temperature to set in their houses," says Millette. "The thermostats are very precise and clear. So in the end, people bought them for the comfort factor." Quebeckers' interest in solar energy is a question of values, and many will be willing to pay the price for something that is in accordance with these values.

The Sun Slowly Rises

What "bothers" Hydro-Québec about solar energy is the fact that everyone seems to believe solar is the best way to save energy. While in fact, energy efficiency measures are affecting Hydro-Québec sales more at the moment. These are certain to have the biggest impact on Hydro-Québec in the medium and long term as well.

When we visited Yves Poissant's home, he invited us into his living room to demonstrate yet another feature of his solar energy "experiment": his windows. A large portion of his investments in solar energy actually involved putting in thicker insulation and double-paned windows, and cutting holes for windows on the south side of his house to allow more sunlight to enter. These features, known as "passive solar heating," do more to make a house energy efficient than a solar panel system.

Contrary to most energy self-generators, who might dabble in self-producing, Yves Poissant is a scientist who scrupulously documents the energy performance of his house. He wants to make sure the calculations of his energy savings are not influenced by either wishful thinking or selective memory.

The savings are, indeed, impressive. "When I compare our house to my neighbours', I see that a house this size, with air conditioning and a pool, normally consumes four times more energy on an annual basis," he says. Poissant even analyzed his electricity generation to single out the savings from the solar energy system from his other energy-saving measures. "I'd say up to 75 percent of the reduction in energy consumption comes from the passive solar measures and measures that are part of the Novoclimat certification (his builder had to conform to a set of standards to ensure energy efficiency). Those are what brings the most savings, not the solar system," says Poisssant.

Poissant's words get to the heart the solar energy matter. If Poissant hadn't worked on the design and layout of the house from the outset, his investment in solar heating wouldn't have paid off. And even at that, Poissant's house is not autonomous in the event of a power outage. "For now, my backup battery is Hydro-Québec," he says. To be fully autonomous, Poissant would have to spend another $12,000 to $15,000 on a battery that could store solar energy and install an emergency transfer switch.

In a way, it's possible Hydro-Québec is overreacting to the rising popularity of solar panels. In spite of the hype, solar energy may never make serious inroads in Quebec. Even in Ontario, the energy self-generation phenomenon is growing more slowly than it probably should be. Starting in 2009, the Ontario government offered a pile of perks to self-generators to encourage the development of distributed solar energy generation, including subsidies and astronomical rates for purchasing back electricity produced through solar energy (Ontario pays up to four times regular electricity rates to buy back solar energy). And even with those perks, Ontario only has 30,000 self-generators using alternative energies, all types combined. Yes, that's 40 times more than the number of self-generators in Quebec, but it only amounts to 0.5 percent of residential customers in Ontario—in spite of the juicy incentives offered by the government, and the progress alternative technologies have made in recent years.

The bottom line is, self-generating electricity is just not everyone's cup of tea. Like gardening, residential solar energy demands a high degree of personal involvement (to plan, install, maintain and manage the equipment, and then to troubleshoot). Growing your own carrots ends up being marginally cheaper than buying them at the grocery store, but 99 percent of consumers still get their carrots from a store because growing them means you have to be ready to get your hands dirty and devote time to a garden. And then there's the fact that, in most climates, you can't have a steady supply of home-grown carrots all year long. Solar heating presents a similar dilemma. Like gardeners, self-generators need to do it for the pleasure of it. And they need a backup source of energy.

Of all the changes Hydro-Québec sees going on "behind the meter," the adoption of energy-saving measures is the most significant. These measures explain why energy consumption has stagnated in Quebec over the last decade. And energy efficiency measures will continue to gain momentum since the government is promoting them so strongly.

Millions of homes in Quebec now have electronic thermostats that allow them to reduce energy consumption by up to 10 percent. And that doesn't include customers who are already taking advantage of government energy efficiency programs. According to TEQ,

Comparison of distributed generation and energy efficiency programs

Program	Number of participants	Energy savings (gigawatt-hours)	Annual energy savings per customer (kWh)
Net Metering (HQ)	716	1.1*	1,500
Rénoclimat (TEQ)	106,562	513.6	4,819
Éconologis (TEQ)	83,132	38.4	438
Novoclimat (TEQ)	42,698	191.9	4,558
Total TEQ	**232,392**	**743.9**	**3,200**

* Based on 1500 kWh of distributed generation per customer

Fig. 5-B: Transition énergétique Québec's (TEQ) energy efficiency programs result in more energy savings than the total generated by distributed energy generators registered with Hydro-Québec's Net Metering program. The comparison is on the basis of self-generators with 10 photovoltaic solar panels that produce 1500 kilowatts of home energy per year.

the Quebec government agency that coordinates and administers them, 232,000 Quebec residences, or 6 percent of Hydro-Québec's residential customers have been using the programs since 2007. The combined savings from these programs are 0.7 terawatt-hours per year, or enough energy to supply a city the size of Sherbrooke, Quebec for four months. What's more, these statistics (Fig. 5-B) only include homes that applied energy efficient measures through official government programs. The real number is certainly much larger. If you take Yves Poissant's annual electricity production from solar energy (about 1500 kilowatt-hours) and multiple it by 716 self-generators, the total would only amount to 1/700th of the savings made annually through TEQ programs.

As we saw in Yves Poissant's solar experiment, energy choices, whether individual or collective, are actually a mix of technical, financial and psychological (or even philosophical) compromises. Solar energy is so popular today that people's decision to adopt it is based strictly on their values, or in the cases of businesses, to enhance their image. Given that other things like batteries, advanced home automation and energy efficiency measures achieve the same objective as solar energy, with less effort and money, will the symbolic power of solar energy end up overriding economic and even environmental logic?

Éric Martel is convinced that the answer to the riddle will come from better defining the objective, as opposed to focusing on a single piece of technology. "Reducing greenhouse gas emissions is also about the energy we *don't* use. Whatever kilowatts Hydro-Québec Production doesn't sell to Hydro-Québec Distribution, at 2.9¢ per kilowatt-hour, it can sell in the U.S. market for 6¢ per kilowatt-hour. In that case Quebeckers will collectively make more money and help the environment at the same time," Martel explains.

In short, Hydro-Québec's next big reservoir of energy might not be a faraway lake or river in Labrador, Quebec's North Shore or Nunavik, but bungalows in Val-d'Or, Montreal North or Alma. Quebec Premier François Legault sees the situation the same way: "We can achieve a lot with energy efficiency alone, and it's the least costly choice at the moment," he said in our interview.

Behind the Meter

To better understand how consumer mentalities are changing, Hydro-Québec reoriented the mission of its Laboratoire des technologies de l'électricité (electrical technologies laboratory, LTÉ) in Shawinigan.

The research centre was created in 1987 to study how electricity was being used on "the other side of the meter," i.e., in Quebec's houses and factories. (Studies at·the IREQ research centre focus strictly on Hydro-Québec's generation and distribution grids.) When LTÉ opened, its main mission was to help industrial customers switch to electric power. Among the centre's achievements, it introduced electric driers to the lumber industry and electronic thermostats in Quebec homes.

LTÉ recently switched its research focus to studying how "participating" customers are affecting the system. "A lot of our research at the moment focuses on behaviour," explains John Gaspo, Director, Energy Technology Laboratory at LTÉ. "For the last three or four years, we've seen customers who are no longer just consumers. They want to get involved, produce their own energy, be autonomous."

LTÉ is casting a wide net, looking at solar energy, energy storage, geothermal energy, but also at how communications systems allow customers to control energy consumption from a distance. The field is known as advanced home automation. The centre is conducting a number of studies on home automation in two "house labs" it built in Shawinigan. From a distance, nothing about the two grey homes looks out of the ordinary, aside from the fact that they are on the grounds of a research centre and that no one lives in them. There are no cars or bikes in the driveways, but the houses are furnished with beds, tables, armchairs and appliances so researchers can calculate the impact of these objects on air circulation and heat production.

Each room has a tripod with a thermometer to measure strata of heat in ten-centimetre slices. Other less-visible sensors measure airflow, heat radiation and heat loss. Researchers want to see what happens to energy consumption when the blinds are shut, or when the garage is heated, and so on. Each house is fitted with two batteries: a Tesla and another built by Hydro-Québec, so researchers can compare the efficiency of both models. The houses also have emergency electric panels so researchers can see what will happen to the automation system during a power outage. Each house has two electricity meters: one exactly the same as what is found in 98 percent of Quebec houses; the other, Hydro-Québec's new "communicating" meter that allows customers to manage their home appliances from a distance, a service that might soon be available to Quebeckers.

"The real value of home automation will be in the ability to manage light bulbs, switches, communicating thermostats and water heaters from a distance, and to combine that with geo-tracking," says Vincent-Michel Duval.

The LTÉ research centre has already come up with some interesting findings. "We know that if you have four communicating thermostats in a house, it's possible to save 1000 watts during peak hours. With 10 communicating thermostats, that number increases to 2000 watts," says Jocelyn Millette, who explains that LTÉ is also studying how to control curtains and lighting from a distance. "There's a lot of talk about the Internet of things and interoperability, but the technology is not there yet," he told us, pointing to a pile of small routers on the counter, near the stove. "Home automation is quite advanced when it comes to security, but the products to control energy on the market don't perform well. And what does exist, can't communicate, so there isn't much potential for interoperability, not yet."

Hydro-Québec is not the only organization whose researchers are studying the possibilities of advanced home automation. Quebec's Université de Sherbrooke has its own advanced home automation centre, and the CanmetENERGY laboratory in Varennes is also very active in the area. Éric Martel is convinced that, in the long run, homeowners will find home automation and energy efficiency to be much more attractive options than photovoltaic solar energy. "Home automation will simplify life and give customers more control, much more than solar panels will. And the lifespan of solar panels is still relatively short."

Aside from the challenge of connecting the parts to make the whole system work efficiently, advanced home automation has to solve another challenge: cyber-security. Communicating thermostats, the Internet of things and the ability to control energy from a distance may be noble goals, but systems need to be secure enough to prevent pirating, ransoming, identity theft, spying and computer hacking. Since 2010, power systems across the planet have seen a spike in hacking, and system managers and governments themselves are increasingly worried. By exponentially multiplying the number of entry points, advanced home automation is making their power systems all the more vulnerable.

"We've learned a lot from the two experimental homes. The main problem is that no one is living in them," explains Jocelyn Millette, who is pushing for Hydro-Québec to adopt a "living lab" approach to its research by getting customers, suppliers and the entire Hydro-Québec division involved. "Whether you call it a living lab, or a customer-based approach, LTÉ can't be the only one doing this research. Hydro-Québec's three business units, TransÉnergie (for transmission), Production and Distribution, will all have to embrace this approach. We need to test new products and procedures, and we need to be able to explore how customers are experimenting with them on their own."

As customers change the way they are using electricity, Hydro-Québec has to know what this will do to the power system.

Chapter Six

Testing the Power System

Some batteries, like cylindrical 1.5-volt AAA batteries, are so small you could mistake them for a throat lozenge. Others, like rectangular 9-volt car batteries, are the size of a (really heavy) lunch box. Then there are the batteries that arrived in the Inuit village of Quaqtaq in July 2018. Residents of the village located on the north point of the Bay of Ungava were probably stunned when a Hydro-Québec barge landed on their shore carrying three 600-kW mega-batteries, each the size of a shipping container weighing 10 tonnes.

Quaqtaq doesn't have a port, or even a quay, so unloading ships is complicated at the best of times. The village's 376 inhabitants have only two physical links with Montreal, 1750 km away: daily shuttles by Air Inuit, and twice-yearly supply deliveries by the shipping company Groupe Desgagnés. In good and bad weather, the ships lower their anchor one kilometre from the shore and unload their cargo by crane onto a barge, which is pushed to shore by a tugboat. A loader then lays down heavy metal sheets that allow it to drive onto the barge to unload it. Emptying a ship is a painstaking operation that requires numerous trips back and forth from shore.

The three mega-batteries that Hydro-Québec delivered to Quaqtaq were an essential step in a pilot project the company has been carrying out there since 2017. Hydro-Québec's objective is to combine the village's diesel generators with new, alternative energy sources. Hydro-Québec installed solar panels in 2017, and batteries in 2018. In 2019, it will be repairing the existing diesel generator. Over the next decade, Hydro-Québec is also hoping to add windmills to the mix.

In hydroelectric jargon, Quaqtaq is what we call an "off-grid system," meaning it isn't linked to Hydro-Québec's main power system. There are 25 off-grid systems in Quebec, including one in each of the province's 14 Inuit villages, others in the North Shore region, and one in the territory of the Attikameks. The biggest one, in the Magdalen Islands, is as large as the 24 other off-grid systems together. If the experiments in Quaqtaq are successful, Hydro-Québec hopes to convert off-grid systems to alternative energy sources at a rate of three per year.

Hydro-Québec has two good reasons for converting these systems: the first is to improve the company's carbon footprint, even though it is already excellent. The second is to reduce the cost of supplying energy to the 35,000 customers in the remote communities. Hydro-Québec is carrying out its most advanced experiments in new energy sources and battery power in Quebec's Far North. The costs of operating diesel generators there are astronomical—up to 60 cents per kilowatt-hour. Since Quebec nationalized its power system, Hydro-Québec has had to sell electricity to all Quebeckers at the same price, even in places where it costs 10 times more, like off-grid systems. The customers of these systems pay an average annual bill of $1140. When the cost of those systems is added up, it amounts to a loss of $200 million, or $6000 per year per customer, for Hydro-Québec.

The idea of replacing part of Hydro-Québec diesel generation with wind or solar energy makes financial sense—unless, of course, some of these grids can be linked to Hydro-Québec's main power system. Hydro-Québec is investing $100 million to build a 220-km underwater power line linking the Magdalen Islands to

Gaspésie, slated to be finished in 2025, and will be investing as much to join the new Romaine Complex to the system when it's complete. But with its extremely cold climate, remote geographic location and general absence of infrastructure, Nunavik poses a different challenge. Hydro-Québec is experimenting with avant-garde technologies in small, costly systems here because alternative energies may pay off right away.

If the experiment in Quaqtaq is successful, Hydro-Québec will expand the project to other off-grid systems in the south. It will start on the other side of Quebec in downtown Lac-Mégantic, which was destroyed in a catastrophic train explosion in 2013, then in the Magdalen Islands. In the new locations, the company hopes to apply the lessons it learns in Nunavik. In Lac-Mégantic Hydro-Québec will be building its first microgrid, a system to be managed semi-autonomously that will combine solar energy, batteries and advanced home automation technology.

The results of these two experiments in the north and in the south will be extremely important for Hydro-Québec's future. The fact is, even though the benefits of new energy sources appear to be obvious, Hydro-Québec can't start implementing them without testing them first. Injecting novel forms of generation (for Quebec) like solar and wind power into the system involves considerable technical challenges. The issue is not how to install solar panels on the roofs of individual homes. It's about feeding entire systems, in part, with new energy sources. Since solar and wind power are by nature intermittent, they change the dynamics in the hydro-electric system. Traditionally, the challenge to power systems has been to handle variations in demand (people use more electricity, for instance, at breakfast and supper time than in the middle of the night). But wind and solar technologies add another, new complication: variations in *generation*. The wind doesn't always blow and the sun doesn't always shine.

The experiment with new energy sources in Quaqtaq is an opportunity for Hydro-Québec to study how to manage solar and

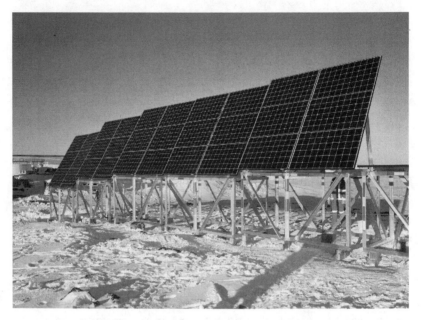

Fig 6-A: Hydro-Québec introduced large-scale solar energy generation in the Inuit village of Quaqtaq on an experimental basis. Depending on the results, Hydro-Québec may build solar panels in other communities in Nunavik, as well as microgrids like in the city of Lac-Mégantic or the Magdelan Islands.

wind power in the system before introducing them on a wide scale. This is a luxury power companies have rarely had. In Germany, California and Hawaii, utilities have introduced massive amounts of solar energy amounting to 7, 10, or even 12 percent of their overall capacity. What they've seen is that, over these thresholds, power systems start having serious difficulty adjusting to variations in generation that can overload circuits, burn lines and lead to brownouts or blackouts.[1] In Quaqtaq, we're not talking about adding 10 or 12 percent of renewable energy, but between 25 and 75 percent depending on weather, and the season. In both Lac-Mégantic and the Magdalen Islands the proportions will be similar, as they will be in other off-grid systems with alternative energies.

"With the creation of microgrids, and the introduction of energy self-generation and batteries, Hydro-Québec is entering a new era of distributed generation," says Vincent-Michel Duval, Engineer in charge of integrating new technologies at Hydro-Québec Distribution. "Because of solar and wind power, batteries and home automation we don't even know what a power system will look like in 50 or 60 years. We might end up not needing some of our generating stations and lines within 20 or 30 years."

Breaking Ground in Quaqtaq

There were a number of reasons Hydro-Québec chose the off-grid system of Quaqtaq for its first pilot project using alternative energies. Hydro-Québec was planning to add an extra generator there in 2018 anyway to make up for projected power shortfalls. There was also space available to install solar panels on the site.

Hydro-Québec is actually testing three things on the houses of Quaqtaq: its first solar park, a new model of mega-batteries and distributed solar energy generation. The goal is to see how much the system can save in diesel and learn how to coordinate diesel generators with variable energy sources in a single system.

Hydro-Québec has many research laboratories working on the various technical challenges in Quaqtaq. The Laboratoire des technologies de l'électricité is developing instrumentation that can be used to gather data from houses and buildings. The IREQ research centre is creating software to coordinate generators with batteries and alternative energy sources. It is also creating computer models that will compare the efficiency of solar and wind energy. IREQ's researchers have already concluded that wind power is more efficient than solar. "Like solar, wind energy is variable. However, the wind always blows, at least a bit, while Nunavik has very little sunshine in the winter," says Richard Lagrange, Director of Off-grid Systems at Hydro-Québec. Experimental projects like Quaqtaq always involve a degree of trial and error, he tell us. For example, his team is studying 100-kilowatt windmills, which are significantly smaller and less powerful than those used in southern Quebec. "In most isolated places, we have to be able to install and maintain the equipment without access to cranes or any type of heavy machinery," he explains.

Hydro-Québec is trying out all sorts of new things in Quaqtaq. Mega-batteries, for example, can be used to "regulate" solar energy produced in the middle of the day, when demand is at its lowest. The batteries regulate generation by distributing energy equally throughout the day. But a mega-battery can also operate as an auxiliary generator, stocking energy that can be used later, during the peak period, if needed. That could help avoid the need to start up another generator or two in order to meet peak demand (in electrical jargon, this is called "peak clipping"). Hydro-Québec is also experimenting with decentralized management: a third of the solar panels in Quaqtaq are on buildings, and two thirds are in the solar park. Hydro-Québec is also trying out a system of solar panels combined with batteries on four houses, a program called Derrière le compteur (Behind the meter).

"You are probably wondering, why only four houses? It's because we are testing four different types of batteries," explains Patrick Labbé, Manager, Innovation, Major Projects and Conversion. "If it pays off, putting solar panels and batteries in new houses in the south could be an interesting avenue to explore for introducing more solar energy."

Hydro-Québec is hoping to cut Quaqtaq's total diesel consumption by 40 percent, which is a lot, given that the village's generator consumes some 800,000 litres of diesel per year, or the equivalent of eight tractor trailers full. But Hydro-Québec has four other objectives: "We have to reduce greenhouse gas emissions, reduce costs, provide reliable solutions, and finally, make sure the community is in agreement about the project," says Richard Lagrange.

Alternative energies certainly look like a smart way to spend less money on diesel fuel. The reality is, the cost of making these types of energy operable in the north is actually so high they may cancel out any savings. "In a context where moving even one employee costs a fortune, and where there are no hotels or restaurants, costs are sky-high. For example, if it turns out we need 15 workers instead of 10, the project might not pay off. So we have to work within the limits of the subsidies available," says Patrick Labbé, who has to calculate the cost of equipment, transport, moving personnel and equipment, building infrastructure, maintenance and repair, etc..

And, of course, the viability of the project is conditional on the community even accepting it in the first place. Hydro-Québec simply can't take this for granted (we explore the topic further in Chapter 15). One might assume the Inuit communities of Quebec would be thrilled to have a way to reduce their diesel consumption by 40 percent, while introducing environmentally friendly solar or wind power. But it's not that simple. "The problem is actually commercial. The Fédération des cooperatives du Nouveau-Québec (FCNQ) supplies the diesel we use in the generators. Their initial reaction to the plan to introduce renewable energy was, 'So what's in it for us? We'll be losing sales,'" explains Richard Lagrange. When he understood that the stakes for the Inuit Community weren't just technical or financial, Lagrange added a specialist in relations with Indigenous communities to his team to make sure he could strike a deal that would satisfy everyone.

The Inuit communities are is not the only ones involved. If the Quaqtaq experiment is conclusive, and the model is used in other Indigenous communities, Lagrange's team will then have to convince Quebec's Régie de l'énergie of its merits. "The Régie will expect us to be able to prove that the solution we are proposing is the best one. We'll have to demonstrate that it will be less expensive than simply linking these communities to the main electricity network," says Vincent Desormeaux, Advisor, Energy Transition and Indigenous Affairs, Off-grid Systems at Hydro-Québec Distribution. "It won't necessarily be a smooth process."

Lac-Mégantic to the Rescue

The lessons learned in Quaqtaq will probably be first applied in Quebec's "south" in the town of Lac-Mégantic, several kilometres from the Maine border.

Six years after a train transporting oil exploded in 2013, reconstruction of the gaping hole that was once the Lac-Mégantic downtown has just begun. Hydro-Québec decided it would use the opportunity to install the country's first microgrid here. A microgrid is actually a cross between distributed generation, an off-grid system and a centralized system. In Lac-Mégantic, Hydro-Québec will install 1000 solar panels on and around roughly 30 buildings. In addition to the 300 kW of power the solar panels will generate, the city will have a 300-kW mega-battery to regulate solar generation as well as smaller batteries—about the size of a dishwasher—in each of the 30 buildings. These buildings will also be equipped with cutting-edge home automation technology and charging stations for electric cars.

"We are starting with Lac-Mégantic because everything there has to be rebuilt from scratch," says, Vincent-Michel Duval Head of Integration of New Technologies at Hydro-Québec Distribution. Lac-Mégantic's microgrid, which will cost $9 million to build, is expected to be up and running by early 2020. It's possible the entire downtown will be energy self-sufficient during certain periods of the day, though exactly when, and to what extent still remains to be seen. "It's an open research project where the local population plays an active role. We want to push home automation technology to the point where residents can control their curtains and lights from a distance. We're also going to see what other kinds of services we can offer. And we'll see how people react to home automation, batteries and intermittent generation," says Vincent-Michel Duval. "The goal is not necessarily to achieve total self-sufficiency, but to figure out how to manage this kind of structure within the main power system."

There are a number of projects like the one in Lac-Mégantic currently being carried out in the United States. The Duke University campus, in North Carolina, operates on a microgrid. An ambitious microgrid project is also being carried out in the Navy Yard of Philadelphia, a waterfront urban development project. When finished, the $4-billion residential and industrial real estate development will have 35 megawatts of self-generation. In New York, the Brooklyn Microgrid links 60 dwellings to produce 1.25 megawatts of solar power that the owners resell to the neighbourhood.[2]

Though inspiring, these experiments won't produce one-size-fits-all solutions for alternative energy supply that can be applied anywhere. On Île d'Orléans, near Quebec City, for example, Hydro-Québec said "no" to a group of citizens who claimed they had the right to form their own autonomous microgrid. The estimated price tag for the project, $100 million, was prohibitive, and the project called for building windmills all over the island. On the other hand, Hydro-Québec said "yes" to a microgrid project in the Magdalen Islands, presently using diesel generators. The reason? The microgrid will reduce overall costs, and even after Hydro-Québec will have built the costly underwater cable to link the islands to the grid, the microgrid will remain useful in reducing costs and provide alternate supply in case the link is broken.

Hydro-Québec's work and research in the fields of solar energy, microgrids and distributed generation are all part of its ongoing research on alternative energy sources. Hydro-Québec has been studying emerging energy sources since the 1970s when the company tested its first windmills in the Magdalen Islands.

No power system today wants to have all its eggs in one basket. There are certainly advantages to operating with a single technology, like hydroelectricity. There aren't many drawbacks to hydraulic energy, either from a technical, ecological (explored in Chapter 13), economic or financial perspective. Quebeckers made a good choice when they decided to develop it: hydroelectricity is

Fig. 6-B: A hydrokinetic turbine about to be submerged in the rapids in front of Old Montreal (above). Wind farm in Cap-Chat (below). Hydro-Québec will continue to experiment with new technologies.

not only the least expensive and most abundant power source on the continent, it's also the greenest.

Yet hydroelectricity does have one flaw: it relies on a pattern of steady rainfall. In the early 2000s, a prolonged period of drought in Northern Quebec had dramatically lowered the water levels of Hydro-Québec's biggest reservoirs. In 2003, Quebec's government almost panicked when the situation made headlines. André Caillé, Hydro-Québec's CEO at the time, tried to be reassuring. "Don't worry. Quebeckers will have enough energy to roast their turkeys next Christmas," he said. But the truth was, Hydro-Québec was anticipating strong growth in energy demand over the medium term and had already launched projects to add 3700 megawatts of wind power and 5000 megawatts of hydraulic power (by building new dams), plus another 1400 megawatts from various contracts with smaller hydroelectric generating stations and biomass cogeneration. In recent years, the Quebec press has criticized Quebec's government for forcing Hydro-Québec to sign external supply contracts, notably for wind power, when it could have just built windmills of its own, certainly at a lower overall cost. Premier François Legault told us in our interview that the government had, indeed, made a mistake with the private windmill contracts. "We looked at whether we could cancel the wind power contracts but that was impossible. Anyway, the fact is, financially, it wouldn't change much. It would cost us as much to cancel the contracts as we would save by cancelling them."

These considerations don't change the fact that from 2003 to 2005, wind power looked like an interesting option for meeting short-term increases in power demand. The same way one might advise an investor to diversify a portfolio to reduce risk and create a buffer during financial downturns, power system managers like to have different energy alternatives at their disposal in case of emergencies. These could either be interconnections with neighbouring systems or backup facilities. Each works like an insurance policy. Big public institutions, like hospitals and airports, have their own emergency generators based on the same logic. Backup systems cost a fortune, but even if Quebec has one of the most reliable systems in the world, Hydro-Québec has to be ready for anything.

Hydro-Québec also has to explore other energy sources in case it ever faces a water shortage. After trying out wind power, the company is now starting to explore solar energy in the form of solar energy parks and self-generation (technically called "distributed generation").

Hydro-Québec will also keep experimenting with new energy forms, like geothermal energy (which harnesses heat from below the earth's surface) or osmotic power (which taps the energy released in the contact of salty seawater and fresh river water). Since Quebec has a surplus of energy at the moment, neither of these options—especially osmotic power—is economically worthwhile. But no one can see the future, and the history of energy has always been marked by unforeseen changes.

In the summer of 2010, tourists visiting Montreal's Old Port were probably puzzled by a large object that looked like an aircraft engine sitting on top of two barges in the middle of the St. Lawrence River. Later in the summer engineers lowered this "engine" 10 metres below the surface of the water. The equipment was actually a hydrokinetic turbine, an underwater windmill that is set into action by the current of the river. It fed the Hydro-Québec grid until 2013. This kind of power at first sounds like an excellent idea for meeting the needs of Quebec's most isolated communities. The problem is, a number of challenges have to be met before it can be used in Quebec's North. How do you repair an underwater turbine running under six metres of ice, during -40° C weather? For that matter, how will Hydro-Québec make sure the ice doesn't damage the apparatus in the first place?

In the meantime, Hydro-Québec is closely watching ongoing research at two universities related to another promising avenue: hydrogen energy. Hydrogen energy makes it possible to stock energy in gas form. Electrolysis breaks up water molecules so hydrogen can be stored in a tank. The hydrogen gas is then combined with oxygen using a special membrane that captures electric energy from the reaction. Quebec, with its bountiful reserves of fresh water and hydroelectricity, is an ideal environment to develop a hydrogen industry—but again, a number of problems must be solved before this technology can be economically feasible. For one, the platinum-based membrane that captures electric energy from the reaction of hydrogen and oxygen is very expensive. The other problem is that the structures necessary to produce, transport and distribute hydrogen energy on a large scale, don't yet exist.

The media have the habit of comparing different types of energy strictly by price. Hydro-Québec's new La Romaine Complex of four dams and power stations (the last of which is to be commissioned in 2021) is a case in point. Its cost price will be 6.5 cents per kilowatt-hour, which is the same as wind power. But price is only one of several factors that determine the value of a given technology. Wind and solar energy are both intermittent sources, so they only work when there is wind or sun. Solar and wind generators seldom operate at more than 35 percent of their capacity, and neither wind nor solar energy come with built-in storage capacity, unlike a hydroelectric dam whose turbines can also be stopped and restarted at the drop of a hat. In other words, the 6.5 cents per kilowatt-hour price of hydroelectricity is, overall, worth more than wind or solar power of the same price because it comes with both generating capacity and potential to be stored. For that reason alone, Hydro-Québec will continue planning hydroelectric dams even if future hydroelectric dam projects have a higher initial price tag than solar or wind farms.

Ahead of the Meter

Alternative energy technologies—whether solar parks, distributed solar energy, windmills, mega-batteries, small domestic batteries or advanced home automation—have their own properties and capacities. But they have one thing in common: when they are added to Hydro-Québec's main grid, it's hard to predict how they will affect it.

The best way to illustrate this impact is what experts in energy circles call the "duck curve." The curve shows how, when solar energy is added to any power system, energy supply and demand end up working against each other.

The expression "duck curve" was coined in California, an early adopter, where solar energy accounts for roughly 10 percent of generation (as opposed to 7 percent in Germany).[3] The duck-shaped curve is a chart that illustrates the disruptive effect of injecting massive solar energy generation into a power system. When solar starts producing in the morning, solar energy supplants the regular energy coming from thermal generating stations. These stations, as a result, have to decrease their production levels to a ridiculously low level. The problem is that just when the supply of solar energy starts to taper off in the afternoon, people return home from work, and overall energy demands in the system increase. This means that the network has to suddenly increase generation by 12,000 or 13,000 megawatts just to supply customers for two or three hours of peak use. Illustration 6-C shows how when joined, the curves (showing the variations in production for each system) make the shape of a duck, hence the name.

For a hydroelectric power system, this kind of fluctuation isn't a big problem because it only takes a few minutes to stop or start a turbine to supply power when it's needed. But it's a different story for almost every power system that relies on conventional thermal (oil, coal or gas) or nuclear energy. It can take hours, or even days to start or stop these generators. This substantially increases the

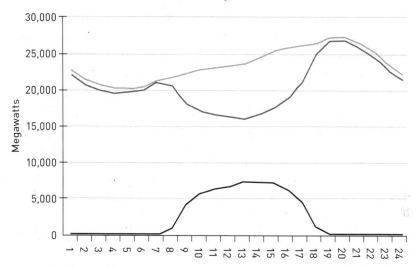

California Duck Curve, October 22, 2016

LEGENDE

━━━━━   Total load

━━━━━   Net load

━━━━━   Solar energy production

Fig. 6-C: The curve, which is literally in the shape of a duck, illustrates the disruptive effect of injecting massive solar energy generation into a power system. The upper line shows how consumption of electricity varies over the day in California. The bottom curve shows how solar generation varies over the day.

operating costs of the system: in California energy generation costs in peak periods have *doubled* since solar energy was introduced in large quantities. So while the sun might be free, using it is still an expensive proposition at the moment.

The duck curve isn't a serious problem when solar generation is limited to about 5 percent of a system's overall mix. Problems start above that ratio. When 10 percent of the system's generation comes from solar, like in California, managing the effects becomes a real challenge.[4] In Quaqtaq's microgrid, where, depending on the season, solar energy represents 25 to 75 percent of generation, the duck curve is more extreme. But variations can be managed because Quebec's off-grid systems have diesel generators that can be started and stopped in just a few seconds. The question is, what would happen in the long term when batteries and distributed solar energy from buildings and houses are added to the system? At the moment, no one has a clear answer.

The effects of variability in generation have been a heated topic of discussion among electric energy specialists for several years. In theory, generation variability should have little effect on Hydro-Québec because it can quickly adapt generation to even the most extreme variations in demand.

But that's in theory. The variability of solar and wind energy makes it difficult to predict generation levels from these sources. What time will the wind start blowing? How long will it blow? How strong will it be? The sun poses the same problem. Close to 8 percent of Hydro-Québec's generating capacity is already subject to unpredictable weather variations. Predictive management—techniques to anticipate fluctuations in generation and act in the minutes *before* problems arise—is now an important issue at Hydro-Québec's IREQ. Back in the 1970s and 1980s, IREQ's main mission was to make high-voltage transmission possible (see Chapter 2). Today, the principal focus of its research is to find a way to manage the system in an energy transition that is forcing Hydro-Québec to add fluctuating and irregular energy sources to its grid. IREQ's researchers are studying how alternative energies disrupt the system, whether its temporary power excesses, dips in voltage or variations in the power.

"Power systems seem abstract to most people. But the truth is, they are living things. Turbines start and stop, people turn up their heating or turn on lights. Demand and generation vary constantly, producing distortions in the electric current. You can see this in houses with poorly designed circuits, where lights dim when the fridge motor kicks in," says Olivier Tremblay, Researcher in charge of power system simulations at IREQ. "The system needs to have built-in protection mechanisms. We need to study how alternative energies interact and how the system reacts to them. If the sun suddenly comes out and a neighbourhood equipped with solar panels starts producing electricity, we don't want to fry TV screens in the next neighbourhood over."

IREQ is carrying on a large-scale research project in order to determine exactly how to avoid problems like this when different technologies are combined. IREQ's Centre de calcul de haute performance (High performance computing centre) has 250 black computers stacked one on top of another, held together with metal brackets. These monitor data coming from one of the fields surrounding IREQ, where Hydro-Québec is building a small-scale replica of a distributed distribution system. It includes houses with solar panels (with and without batteries), a small solar park, a small wind farm, mega-batteries, houses with electric cars and

Fig. 6-D: System simulation at IREQ. The centre is conducting research to figure out how to predict the system's behaviour when variable and distributed generation sources are added.

unidirectional and bidirectional charging stations (programmable and not). "We want to be able to know our computer models work in real life," says Olivier Tremblay.

Aside from the problems of variability and unpredictability inherent to solar and wind energies, solar poses another problem: if self-generators have the ability to inject their surpluses into the system, the electrical current in the system will have to be able to move in both directions. This phenomenon didn't affect the system at the time of writing because there were only 716 self-generators in Quebec. But it will be a different story when entire streets and neighbourhoods generate energy. The system was only designed to work in one direction. It would be like if residents of a city all started collecting rainwater and sending it back into the city's water system through their taps. This would no doubt cause backups and break pipes. The same sort of thing will likely happen with electricity, though at the speed of light. And that's what Olivier Tremblay and his colleagues at IREQ are studying right now.

IREQ developed a real-time system simulation technology called HyperSim, which operates at the same speed as the system itself—the speed of light—and makes it possible to test different maneuvers future system managers could make to regulate the system. Which circuit breaker will need to be cut in what situation? How will this affect the transformers on posts 10 metres in the air? The system can even simulate the functions of an entire wind farm. "The goal is to be able to model unwanted phenomena so we can either eliminate them or come up with set responses to them. When computers detect anomalies, they have to be able to analyze them within milliseconds to correct them properly," explains Olivier Tremblay, who is presently researching a way to amplify the electronic signals HyperSim sends so they can be reproduced in a real system. "We're getting there."

Chapter Seven

The Smart Bill is Coming

Jean-Benoît grew up in Sherbrooke, Quebec, which was for many years known as "The Electric City." On his way to school every morning he crossed the Magog River Gorge on the Hubert-C.-Cabana Bridge, which offers a picturesque view of Sherbrooke's historic Frontenac Power Station, built in 1888. Still operating 131 years later, the station—the oldest in Quebec—still produces 2.2 megawatts of power.

There isn't, however, any mention of the Frontenac Power Station on Hydro-Québec's website, nor of the eight other generating stations operating in Sherbrooke. That's because Sherbrooke residents' electricity bills come from their own para-municipal company, Hydro-Sherbrooke, which has been managing the city's power network since 1908. It's Hydro-Sherbrooke, not Hydro-Québec, that takes care of the manholes, hydro poles and transformers in Sherbrooke, and Hydro-Sherbrooke's logo that appears on maintenance workers' helmets and trucks.

"We have the second-biggest power system in Quebec after Hydro-Québec!" jokes Christian Laprise, Director of Hydro-Sherbrooke. With its 82,700 customers and 150 employees, Hydro-Sherbrooke is definitely the biggest of the 10 para-municipal electric utilities and distribution cooperatives that remained after Quebec nationalized electricity in 1963 (see Chapter 3). In 2018,

Fig. 7-A: The Frontenac generating station in the Magog River Gorge in Sherbrooke. Property of Hydro-Sherbrooke, it is the oldest generating station in Quebec, operating since 1888, and produces 2.2 megawatts of power.

Hydro-Sherbrooke generated $22 million in profits, all of which went back into the city's coffers.

Hydro-Sherbrooke may be a vestige of the past, but it certainly isn't a dinosaur. On the contrary, the company is avant-garde in load management, particularly in the way it deals with the challenge of peak period demand. Hydro-Sherbrooke has made ample use of the resources its local network of universities and numerous engineering firms affords. The company buys its electricity from Hydro-Québec—for $120 million per year—then uses a series of original techniques to control demand in order to reduce its costs, boosting profitability in the process.

Hydro-Québec could learn a thing or two from Hydro-Sherbrooke.

Starting in 1987, Hydro-Sherbrooke introduced a voluntary program for customers to shut off their water heaters for a few hours during peak demand periods. Hydro-Sherbrooke also spread dual-energy heating (combining electric heating with oil or natural gas) to 5 percent of its residential customers (double the rate of Hydro-Québec). It developed more sophisticated methods for remotely controlling electricity consumption. And it introduced a system for starting up the emergency generators of its biggest institutional customers, including Sherbrooke's universities and research centres.

"Combined, these measures allow us to reduce our power output by 70 megawatts," explains Laprise. "Peak demand in winter for us is 528 megawatts. Without these measures, it would be much higher, close to 600 megawatts. Right now we're looking for a way to reduce demand by another 15 megawatts."

Like Hydro-Québec, Hydro-Sherbrooke has always had to keep up with power deficits: the difference being that unlike Hydro-Québec, since 1930, Hydro-Sherbrooke doesn't have the possibility of building new dams to meet growing needs. The company has always had to work with what it has. In its early years it bought extra power from Shawinigan Water and Power. Now it manages the peak periods of demand in a way that doesn't require buying extra power. Since Hydro-Québec's affordable option for boosting generation have all but disappeared, the government-owned monopoly has something to learn from the expertise Sherbrooke's "baby Hydro" has developed in managing power shortfalls.

It sounds completely paradoxical: demand for power in Quebec is constantly increasing even though energy consumption hasn't increased in 10 years (explained in Chapter 4). Even if Quebeckers used all the energy efficiency measures available to cut their electricity bill (effectively reducing energy consumption), during peak hours, they still plug more electric devices in than ever, whether at home or at work. Because of this increased demand at peak periods, requirements for power (in watts) are growing steadily (at the average rate of 150 megawatts per year, in fact), while overall demand for energy (the number of kilowatt-hours you end up paying for) stays the same. For Hydro-Québec, the idea of keeping pace with increasing power demand by building more dams and adding more power lines and transformers, isn't really an option anymore. The really "profitable" sites for building dams are getting fewer and farther away. Increasing environmental and social acceptability standards are now adding new constraints Hydro-Québec didn't have before, even for small projects—and for entirely valid reasons.

The one option Hydro-Québec has is to better manage power demand by following the example Hydro-Sherbrooke set 30 years ago. Hydro-Québec actually has an unofficial target of reducing power demand by 5000 megawatts. And though it doesn't advertise this figure, decisions about future investments, rate programs and new services are all made with the goal of reducing peak demand in mind. Hydro-Québec is entirely rethinking its relationship with all its customers, who are principally residential. The big challenge will be to figure out if, like Sherbrookers, Quebeckers on the whole will be able and willing to think about electricity as a limited resource.

New Rates

Hydro-Québec has made little effort to get its four million residential subscribers to adjust their consumption to help it manage its power deficit. Residential customers represent 95 percent of Hydro-Québec's clientele. But Hydro-Québec decided to start with its 300,000 industrial and commercial customers. In a way it made sense: there are fewer of them, but they account for 50 percent of overall demand.

Hydro-Québec's non-residential customers pay some 50 different rates in a convoluted billing system that has one unifying feature: they are billed separately for energy and power. Hydro-Sherbrooke, itself one of Hydro-Québec's biggest customers, pays one fee for the energy it uses in megawatt-hours, and another for the power it uses in megawatts. The strategy for managing energy use during peak periods is to offer incentives (like lowering the total amount of the hydro bill) to organizations to encourage them to shift excess use of power outside of the peak demand periods.

This double billing system that separates energy from power has the advantage of creating incentives for big customers to control their power demands on Hydro-Québec: customers respect certain self-imposed limits to reduce their bill. That, in turn, helps Hydro-Québec control its own costs.

Certain consumers and large industrial customers even participate in what are known as "interruptible" programs, where Hydro-Québec actually cuts their service during peak periods (the customers are, of course, warned in advance). Ski resorts, for instance, voluntarily do this and Hydro-Québec credits them 70 cents per unused kilowatt-hour. "When it's very cold and there aren't that many skiers anyway, it makes more financial sense for ski resorts to close. Hydro-Québec could have a similar program for schools, especially in the evenings," says Vincent-Michel Duval, an engineer in charge of integrating new technologies at Hydro-Québec Distribution. In all, Hydro-Québec is able to save 900 megawatts by applying these measures. Meanwhile, it's studying other scenarios for interrupting power to institutional and commercial customers. But Hydro-Québec hopes to do much more in the long run, following Hydro-Sherbrooke's example. The 900 megawatts Hydro-Québec saves only represent 2 percent of its total capacity: all things considered, Hydro-Sherbooke's measures allow it to save something in the range of 12 percent of its capacity.

The solution would obviously be for Hydro-Québec to get its four million residential customers on board, particularly since Quebeckers are one of the most energy-guzzling populations on earth.

But before looking at what Hydro-Québec's plans are, it's important to understand how the "Rate D" works ("D" for domestic). It's actually an amalgamation of three different rates: two for energy and one for capacity.

The first rate on Hydro-Québec's residential bill (Fig. 7-B) is the base rate, 5.91 cents per kilowatt-hour. This rate applies to the first 40 kilowatt-hours customers use per day, which amounts to 1200 kilowatt-hours in a 30-day month. When we say Hydro-Québec charges the lowest rates on the continent, it's this first 40 kilowatt-hour block we are referring to. About 20 percent of Hydro-Québec's residential customers never exceed the 40 kilo-

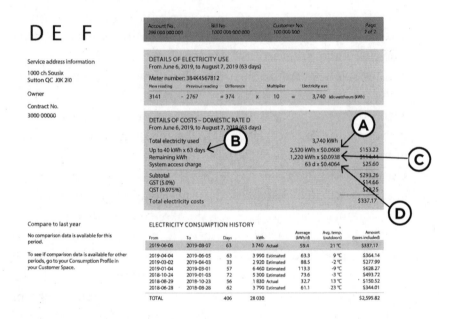

A: Rate for first tier: 6.08 ¢/kWh
B: Daily limit of the first tier: 40kWh-day starting in 2019
C: Rate for second tier: 9.38 ¢/kWh
D: System access charge: 40.64¢/day

* This bill is in accordance with Hydro-Québec business practices at the beginning 2018.

Fig. 7-B: Hydro-Québec's bill. There isn't one residential rate but actually two or three. A is the first rate block. B is the daily limit of the first block. C is the second rate block. And D is the fixed service fee (per day).

watt-hour threshold. The size of this first block has been increased considerably in recent years, from 30 kWh per day in 2016 to 40 in 2019. "Smaller consumers had been asking for this increase for years," explains Marc-Antoine Pouliot, Manager, External Relations Strategy at Hydro-Québec Distribution.

A customer who uses more than 40 kilowatt-hours per month will pay a second rate 50 percent higher, or 9.12 cents for the additional kilowatt-hours. For a household that consumes 1000 kWh per month, the average rate paid by Hydro-Québec customers in 2018 was 7.13 cents per kilowatt-hour. High-consuming customers pay close to 9 cents per kWh. A household that consumes 3000 kWh per month will pay 8.37 cents per kilowatt-hour.[1]

The third rate on the bill is called the "System access charge." It's a fixed service fee of 40.64 cents per day (about $14 per month) that customers all pay even if their consumption is zero. (This can seem steep, but in Ontario, the fee is double and varies according to whether the customer lives in a city, suburb or rural zone.) Hydro-Québec's system access fee goes toward operation costs of the facilities it has to build to meet potential demand. The fee is also the reason Hydro-Québec rates, overall, are so low: everyone would universally have to pay a higher rate per kilowatt-hour if facility operation costs were included in it.

In 2016, Hydro-Québec tried to tackle its power shortfall with a solution that was inspired by Hydro-Sherbrooke: offer volunteer customers the possibility of intermittently cutting power to their water heaters. Hydro-Sherbrooke started doing this 30 years ago. Hydro-Québec's program would have allowed it to cut the electric supply to hot water heaters (of customers who had signed up for the program) during critical winter hours, for periods of no more than five hours at a time, up to a maximum of 300 hours per year.

Unfortunately, Quebec's Institut national de santé publique (national public health institute) killed the program by announcing there was a risk (very slight) that the power cuts could cause Legionnaire's disease. The program was shelved. Hydro-

Sherbrooke also suspended its program until manufacturers come up with a water heater that eliminates this public health risk.

In 2018, Hydro-Québec went back to Quebec's Régie de l'énergie with a new strategy for controlling energy use during peak periods. No more water heaters this time: Hydro-Québec proposed two new voluntary "dynamic pricing" models instead that will take effect in 2019. In both cases, a special (higher) rate will apply during specific hours of the winter peak periods when Hydro-Québec is having difficulty meeting overall power demand and customers will be rewarded for deferring use outside of peak periods.

In the first pricing model, called the Winter Credit Option, the rates for specific blocks of power will remain the same but customers will agree to reduce consumption during peak hours for a total of roughly 100 hours over the course of the winter. A few hours before the peak period, customers will receive a text or email alert. If they manage to reduce consumption during this time, Hydro-Québec will credit them 50 cents for every kilowatt-hour of energy *not* consumed. "The calculation will be based on average consumption during the days and hours preceding the call," explains Marc-André Pouliot. "Fifty cents might seem like a big incentive but it's actually the amount we would pay during the peak period if we had to buy extra power on the energy market."

The other option, Rate Flex D, will let customers take advantage of a 2 cent per kilowatt-hour rebate from December 1 to March 31. However, during the 100 hours of the critical peak period, Hydro-Québec will apply a dissuasive rate of 50 cents per kilowatt-hour, or 5 times the usual price for the second block of the domestic rate. "It's really a limited number of hours. Customers will be notified the day before by email or a mobile application so they can see how much energy they are consuming and reduce during the peak rate period. They will also be able to pull out of the program at any given time," says Marc-André Pouliot.[2]

Energy specialists openly wonder whether Hydro-Québec shouldn't simply charge residential customers for power (i.e., for the kilowatts they use, not the kilowatt-hours), like the company does for its commercial, institutional and industrial clientele. It would be easy to do now that 98 percent of Quebec households are using the new smart meters.

Italy's national electricity utility, ENEL developed technology to do this and used it in 27 million residences between 2000 and 2005. "Italians have a fee system according to blocks of 500 watts. Such a system would have the advantage of translating into a rate the exact amount of power each customer actually draws from the system," explains Pierre-Olivier Pineau, Research Chair in Energy Sector Management at Montreal's HEC business school, who has long argued in favour of reforming Hydro-Québec's rates. The Quebec households that heat with electricity use on average 7000 watts, but this can easily rise to 13,000 in peak periods when everything is turned on. Some of Hydro-Québec's biggest residential customers easily use more than 30,000 watts and up to 40,000 during peak periods: 100,000 of those customers adds up to several thousand megawatts of extra capacity Hydro-Québec needs to have on hand—the equivalent of one very large dam. "The advantage of this system is that it would send a clear price signal to consumers showing them what it actually costs to supply their energy."

Christian Laprise, Director of Hydro-Sherbrooke, is not sure it would be fair to go that far, since heating is an essential need for which, in Quebec, electricity is almost always required. "We have, in a way, socialized the cost of electricity. It's part of a social pact," he says. "In practice, that means any effort Hydro-Québec makes to control power demand has built-in limits. We are not going to ask people to stop heating their homes. The power deficit is not a problem to be taken lightly, but it's a cost we have to assume collectively."

New Services

Even with its voluntary dynamic pricing programs, Hydro-Québec is not prepared to go as far as someone like Pierre-Olivier Pineau wishes it would. However, Hydro-Québec is considering new services that could help it get there.

The new programs mainly target heating, which accounts for more than half of overall household energy consumption in Quebec and makes enormous demands on power supply. "If we lower thermostats by 2 degrees during peak periods, we won't affect customers' comfort but the savings for the system will be significant. It follows the same logic as interruptible hot water heaters," says Marc-Antoine Pouliot. "We will be looking at controlling electric car charging the same way in a few years."

One of the avenues Hydro-Québec is considering is a residential battery system, possibly coupled with solar panels, along the same lines as the system Hydro-Québec is using in Quaqtaq (described in Chapter 6). Batteries could serve as backups for the whole electrical system of houses, including heating, water heating and lighting. "As far as we are concerned, the best scenario would be a combination of batteries and home automation," says Michel-Vincent Duval, Head of Integration of New Technologies for Hydro-Québec Distribution.

These solutions are not the stuff of science fiction. In Vermont, Green Mountain Power, owned by Quebec's natural gas supplier, Energir, introduced a program to rent Tesla PowerWall 2.0 batteries to customers for $15 per month. It's a niche program for rural customers who are having problems with supply or reliability, but it could certainly be used on a wider scale. "In 10 years, Hydro-Québec might be able to install solar panels coupled with batteries that will make it possible for customers to manage peak periods on their own, produce their own energy and inject extra energy back into the system," explains Karim Zaghib, General Manager of the Centre of Excellence in Transportation

Electrification and Energy Storage, or CE-ETSÉ, a business incubator that grew out of IREQ.

Though storing energy seems like an "obvious" solution to Hydro-Québec's power shortfall, it's not as self-evident as one might think. "Batteries are expensive, so you have to ask what that energy will be used for in the end. If it's for supplying electricity, then batteries are a good solution. But if the end goal is heating, it's actually less expensive to store energy in the form of heat, inside people's homes," explains Jocelyn Millette, Researcher at LTÉ in Shawinigan, which is studying customer relations and new energy management processes. "What we think of as a battery isn't limited to electrochemical processes. A water heater or a well-insulated house accumulates energy, so it's a battery too, and we can work with that."

The question of what could function as a battery prompted LTÉ to study a new technique called "preheating." It's exactly the opposite of what Hydro-Québec has been recommending customers to do for years. Until now, customers have been advised to turn the heat down at night when they are sleeping, and during the day when they are out of the house, and crank it up when they get up in the morning and when they get home from work. The problem with this old advice is that it just ends up increasing peak power demand. The logic of preheating is the opposite: heat houses up a couple of hours before people get up, then lower the temperature during the morning peak period; turn it up a couple of hours before they return home, and then turn it down again when they arrive (again, during peak period). This could be done with remote heating technology when customers are out of the house. Together, these measures would reduce power demand at times when most people are taking showers and making breakfast (in the morning) or giving children baths and cooking dinner (at night).

"We tested this technique in 42 Hydro-Québec buildings and asked occupants what they thought. No one noticed any discomfort, and by reducing power demand in peak periods, Hydro-Québec saved $1.7 million dollars," says Jocelyn Millette. Based on its two experimental houses, LTÉ estimates Hydro-Québec would save between 1000 and 2000 watts of power by using the right type of thermostat. Multiply this by four million customers, and the savings would be in turbines and hydro dams Hydro-Québec would no longer have to build to meet rising power demands.

Hydro-Sherbrooke's Christian Laprise is also considering this approach. "It means customers are more comfortable in the morning, because the house temperature is high, and the radiating temperature from walls and furniture just adds to that. The system could work with all different types of heating. It would probably be less expensive than introducing batteries, overall, but we need to calculate how much it would cost to implement."

A system for controlling house heating remotely already exists in Quebec. It's called "dual-energy" heating and is used by 5 percent of residential customers at Hydro-Sherbrooke and 2 percent at Hydro-Québec. Dual-energy heating works as follows: houses have two furnaces, one electric and the other oil or natural gas. During peak demand periods, the electric heating is turned off and the oil or natural gas furnaces kick in. There are just two obstacles to making dual energy work everywhere: first, not everyone has the means to buy two furnaces. Second, dual energy only works with central heating systems and the vast majority of houses in Quebec use electric wall heaters. By comparison, a system for controlling heating remotely would not be affected by either of these problems.

LTÉ's experiments in preheating and home automation are taking Hydro-Québec far beyond differential rate programs. In fact, the lab is now considering services that won't start and stop at delivering electricity, but will actually manage consumption "behind" the meter, inside customers' homes.

Hydro-Québec started experimenting in this area in early 2018 with a pilot project in which it installed 10 "communicating thermostats" in the homes of 50 employees. The thermostats come with a specially designed application that allows them to "speak." Jonathan Côté, Hydro-Québec spokesperson and a self-described "techno geek," insisted on being one of the 50 guinea pigs for the experiment. "Hydro-Québec technicians installed communicating thermostats that I could control individually from my phone. I programmed each one and watched the energy consumption in each room. I was in the United States at the time but I knew when my cleaning lady was there because I "saw" her turn up the heat. The application even allowed me to budget. It said "If you lower the temperature by 2 degrees, you will save this much." The app could even calculate the "ghost charge" of my appliances when they were turned off."

This new application is at the heart of a pilot project Hydro-Québec developed to test new advanced home automation services in 400 households. The product, which doesn't have a name yet, will make it possible to control, remotely, not just communicating thermostats, but lights, curtains and other electric devices.

The new service will also solve one of the big problems Hydro-Québec faces in applying differentiated rates: most customers really don't want to figure out the difference between power and energy, or how to manage their energy consumption by time of day. The advantage of a remote home automation system is that Hydro-Québec could do the managing for its customers and offer more comfort through features like temperature control in exchange for customers using less energy at certain periods.

Meanwhile, Hydro-Québec's President and CEO Éric Martel is trying to evaluate the commercial potential of the experiments Hydro-Québec is carrying out in its "smart houses" in Shawinigan. "It's clear we can now do something with energy efficiency and home automation that we couldn't do 10 years ago. Digitalization and smartphones are making it possible to optimize energy control from a distance. We are testing products that will allow customers to save money and reduce the system load in peak periods by 2, 3 or 5 percent at the same time."

"Basically, we want to offer an interactive home automation service where customers can just pay for the comfort and flexibility they want," explains Jocelyn Millette. Hydro-Québec's home automation division has taken inspiration from practices other electricity companies have already adopted. France's national electricity company, Électricité de France (EDF), for example, developed a home automation division called Sowee that allows customers to remotely control charging stations for car batteries, and to adjust water heaters and furnaces to changing weather conditions. In the UK, British Gas created Hive.

Advanced home automation services will be a new page in the history of Hydro-Québec Distribution. "The goal is to get over the idea that electricity is a cheap commodity," says Millette. "If Hydro-Québec plans on doubling sales, it has to break out of the old paradigm and offer new services."

Chapter Eight

Creating Demand

"**W**elcome to the cloud," Fabrice Fossaert, Manager of the OVH data centre, in Beauharnois, told us after we had passed through the company's heavy security system. OVH is the acronym for the French "On Vous Héberge" (We are your host). That about covers it: the sole purpose of this French multinational company is to host a phenomenal quantity of websites. In 2013, OVH bought a recently shuttered, 43,000 square-meter aluminum smelter just 300 metres from Hydro-Québec's Beauharnois generating station west of Montreal. It then transformed the former plant into one of the biggest data centres in the world. Strolling down corridors, we felt like we were on the set of *The Matrix*, watching some 360,000 servers, crisscrossed with red and blue cooling wires, blink in ghostly silence. "A server here is activated just about every time you do a Google search," explained Fossaert. "Almost all the data that computers, iPads, smartphones and televisions send to the cloud is actually stored, physically, in a data centre like this."

There are currently 46 data centres like OVH operating in Quebec, including centres for Amazon web Services, Google, Microsoft, IBM and Salesforce. That number doesn't include the several server farms in the province that mine crypto-currency. Together with electrifying transportation, data centres are at the heart of Hydro-Québec's strategy to increase local demand and

Fig. 8-A: The OVH data centre is a gigantic server farm (roughly 360,000 servers) housed in a former aluminum refinery in Beauharnois, west of Montreal. The French company was looking for a place to operate a centre in Quebec because of the province's low electricity rates but also because of its cold climate, which allows it to save on air conditioning.

sell energy surpluses it now has because of deindustrialization and the wide-scale adoption of energy efficiency measures.

When talking about Hydro-Québec's notorious energy surpluses, it's important to understand that there are actually three different *types* of energy surplus.

The first type is the "reserve." This is by far the biggest of the three and exists simply because the law requires it. To ensure the power system has sufficient supply in case of a prolonged drought, Hydro-Québec must keep a big enough water reserve to generate 64 TWh of energy over two years, and 98 TWh over four years. So most of the noticeable energy surpluses in Quebec's biggest reservoirs—amounting to many metres deep of water spread over many square kilometres—can't be touched.

The two other smaller surpluses are those of Hydro-Québec Distribution and Hydro-Québec Production. These two business units of Hydro-Québec, which have separate administrations, were created (along with TransÉnergie, the transmission unit) in 1997 to meet the regulations of the U.S. energy market. In practice, this means each division has to manage its own energy surplus.

Using these surpluses isn't as easy as turning on the tap.

Hydro-Québec Distribution's surplus is the product of Quebec government's pricing policies. Hydro-Québec Production has to guarantee that Hydro-Québec Distribution has access to what's called a "heritage pool" of 165 terawatt-hours of energy per year, which it then sells to Quebeckers at the basic rate of 2.9¢ per kilowatt-hour. The heritage pool is supposed to hold enough energy to meet 90 percent of Quebec's needs. (At present, it actually meets about 97 percent of the overall demand.) However, since the Quebec government forced Hydro-Québec Distribution to sign 60 additional supply contracts (mostly for wind power, but also with small biomass generating stations), Hydro-Québec Distribution has 8 to 10 TWh of extra, unused heritage pool power per year. So Hydro-Québec Distribution is looking for ways to sell this extra energy on the Quebec market.

The third surplus is that of Hydro-Québec Production. The actual amount of this surplus is classified information. While it's public knowledge that Hydro-Québec Production imports 31 TWh per year from Churchill Falls in Newfoundland and exports about the same amount, and that Hydro-Québec Production must guarantee Hydro-Québec Distribution its "heritage pool" of 165 TWh of energy, no one knows how much extra energy Hydro-Québec Production actually produces. Hydro-Québec refuses to divulge this information claiming that it needs to keep it secret to preserve its competitive advantage in export markets. However, the company has already stated it would be able to produce enough energy for two other contracts the size of the one it signed for Massachusetts in 2018 (for 9.45 TWh of electricity, slated for operation in 2022).

We can deduce from this that Hydro-Québec Production's surplus ranges, conservatively, from 20 to 30 TWh. Most of this energy is earmarked for exports markets (see Chapter 9).

Exporting energy is not the only way for Hydro-Québec to sell its surplus energy. The company would like Quebec to develop new sectors of economic activity with high electrical consumption. That approach would have the added advantage of reinforcing Hydro-Québec's mandate to contribute to Quebec's economic development. Potential new customers could include more data centres like OVH, but also electrified transport (cars, buses, delivery trucks), greenhouse agriculture and hydrogen energy production. Hydro-Québec Distribution would also be happy to see new aluminum smelters or mines open: both are high energy consumers. Hydro-Québec wants large customers who buy power by the megawatt, and energy by the terawatt-hour. "In the present context, where demand has been stable for the last 10 years, we have to increase demand inside Quebec. To do that, we have to attract new customers. We can't just increase revenues by hiking rates," says Éric Martel.

The interests of Hydro-Québec and the Quebec government are perfectly aligned. Hydro-Québec has the unconditional support of its single shareholder to develop internal markets. To find new internal markets, the Quebec government is starting with its own networks, like Investissement Québec (a company established in 1998 to promote foreign and local investment in Quebec) and the Quebec pension fund manager, the Caisse de depot et placement. Quebeckers can reasonably expect sales of surplus megawatts and terawatt-hours to translate into new jobs, new economic activity and tax revenues. The electrification of transport would even allow Quebec to kill three birds with one stone: reduce oil imports, reduce GHG emissions from oil and gas and increase revenues from electricity.

However, finding the right way to do this will take years. Take the case of electrified transport. Improving the quality of batteries will make widespread adoption happen faster. In Quebec there are now 40,000 electric or hybrid cars on the roads. That number could climb to a million cars by 2030—a fifth of the cars in the province. A million electric cars require three terawatt-hours of electricity to run them, or 2 percent of Hydro-Québec's overall demand. This would not be an earth-shattering change, but if it happened Quebec would import $2 billion less oil per year. In a generation or two, if all the vehicles on Quebec roads were electrified, these numbers would multiply by five.

Yet it's hard to predict the future of a market economy. Ten years ago, the outlook for the pulp and paper industry was grim, but no one imagined it would decline as quickly as it did. Similarly, the explosion in the use of smartphones has multiplied needs in data management in a way no one saw coming. Ten years ago, digital currencies didn't even exist. And that's not even taking into consideration the changes in values that are happening in Quebec, like elsewhere. Who would have predicted that environmental values would strike the popular imagination to the point that "environmentally friendly production" could be a sales argument for practically any Quebec industry, since most of them operate with hydroelectric power? Everyone is in the market for carbon neutral products, "green" manufacturing processes and sustainable services—even to the point of promoting environmentally friendly data processing.

While the government can create and implement policies, energy choice in a free market is a complex issue. Nevertheless, the ever-increasing effort to reduce the environmental footprint of economic activity has become one of Quebec's greatest trump cards, one that could create opportunities no one has yet imagined. But this phenomenon will also force Quebec to make some difficult choices. That's exactly what happened in 2018, when crypto-currency companies from across the planet suddenly got wind of cheap and abundant electricity in Quebec.

It Never Rains But It Pours

Hydro-Québec first identified data centres like OVH as an interesting avenue for growth in 2010. A big centre like OVH requires up to 40 megawatts of power to operate, or as much as an average sized factory, and creates 120 jobs.

"Electricity is by far our biggest cost," explained Fabrice Fossaert. Quebec is an interesting destination for data centres, not just because of its affordable electricity, but because of its Nordic climate. Operating computers produces so much heat that OVH needs to constantly cool them down to keep them between 16 and 24 °C. "At OVH we built our own cooling system but Quebec's cold climate means we can do without air conditioning many months of the year," Fossaert told us.

Data centres are multiplying in pace with the exponential growth of the web economy. OVH is adding 200 new servers per week at the Beauharnois location (by 2019 the France-based company had 27 data centres in 19 countries). The famous GAFAM companies—tech giants Google, Apple, Facebook, Amazon and Microsoft—are all on their way to Quebec. "The biggest players certainly know us. They are looking at how much they would save in electricity by moving here and how they would benefit from the green branding of using hydroelectricity," says David Vincent, Director, Business Development and Sales at Hydro-Québec.

Then suddenly, in 2017, Quebec became a victim of its own success. Hydro-Québec had launched a big offensive to entice data centres from around the world to come build operations in Quebec that year. But it was also the year of the Bitcoin bubble. After this cyber-currency was introduced to the futures market, Bitcoin's value reached unprecedented heights: between May and December 2017, its exchange rate in U.S. dollars rose from $1200 to $19,200. The sudden increase set off a fury of speculation. Hydro-Québec was swamped with hookup requests from server farms set on moving to Quebec to produce—or, in computer jargon, "mine"—more crypto-currencies.

Crypto-currencies are virtual currencies that don't have a central bank, governor, or so much as a central phone system. They exist because of a self-regulating community of users. There are some 1600 different types of crypto-currency, or nine times more than the 180 national currencies recognized by the United Nations. The most famous of them, Bitcoin, was created in 2009 by a group of people who remain unknown to this day. Bitcoin was used mostly in black market activities until it earned some recent credibility by being introduced into futures markets. But like all cyber-currencies, Bitcoin developed in a complete regulatory vacuum. No one is accountable for its existence and there is no formal control over its circulation. Not surprisingly, almost all central banks issued formal warnings to their governments and investors to be wary of crypto-currency.

If the whole issue sounds overly esoteric, that's because it is. Crypto-currencies owe their existence to a computer technology called block chain that makes it possible to identify all the transactions linked to each Bitcoin and each parcel of Bitcoin. Block chain is, in fact, nothing more than an extremely complex system for solving mathematical and computer code puzzles. Server farms devoted to mining crypto-currency only do one thing: make infinite computer calculations. The only thing "policing" the system and balancing supply with demand is the fact that the more servers there are mining Bitcoin, the more complex these puzzles become and the less profitable it is to mine them. Getting there requires unimaginable quantities of energy and computer hardware. It's easy to see why crypto-currency farms want to be in Quebec, but it's also easy to imagine the level of risk Hydro-Québec would assume by jumping head first into the crypto-currency industry.

Curiously, it took the crypto-mining companies of the world until early 2018 to get wind of the favourable operating conditions in Quebec. Then they started filing hook-up requests with Hydro-

Québec like crazy. When Hydro-Québec added up the power supply of the combined requests, its managers just about fell out of their chairs. Quebec's data centre sector presently uses a total of 81 megawatts of power. The cyber-currency mining companies, together, wanted 18,000 megawatts, or *220 times* more power than is being used by data centres. Even the smallest cryptocurrency projects were asking for 20, 30 or 60 megawatts of power. Many wanted between 100 and 200 megawatts, not to mention one massive request for 2000 megawatts. Hydro-Québec was in a situation it had never experienced. Montreal's Bell Centre, where the Montreal Canadiens play, requires about 5 megawatts of power to operate; the downtown shopping and office complex Place Ville Marie requires 20 megawatts; the city of Sherbrooke in its entirety requires 528 megawatts of power; and Quebec's nine aluminum plants, together require 2700 megawatts. This was still a fraction of what, together, the crypto-mining companies were asking for. In Iceland, which has been literally overrun with crypto-mining companies in the past few years, server farms were using more power than the entire residential sector of the country.

Hydro-Québec is always happy to receive large-scale requests for power. But a "big" project is normally something that takes years to get off the ground and gives the company time to get its ducks in a row before delivering. In this case, Hydro-Québec was being asked to supply 18,000 megawatts in one shot, practically overnight. Hydro-Québec, which in principle, is obliged to meet all requests for power, hit the brakes this time. The company just didn't have the capacity to generate and transmit this amount of power. Had it tried, it would have put Quebec's entire power system at risk.

For its part, the Quebec government was concerned about Hydro-Québec getting into a situation where it had to supply so many megawatts to an industry that was not just highly speculative, but also completely unregulated, not to mention dominated by questionable stakeholders, mostly from Russia and China. "It's very risky. The people behind that industry can come in one day and leave the next. It's highly volatile. If the price of crypto-currency falls, they'll be gone," explained David Vincent. For that matter, in 2018, the government of Iceland decided to curtail any further development of crypto-currency: it was just too risky.

In June 2018, Hydro-Québec decided to ask Quebec's Régie de l'énégie for permission to impose a temporary "dissuasive" rate on the crypto-currency industry, and wait for the excitement to cool down. The Régie organized a special hearing. Normally, the Régie's hearings only draw a small number of representatives of citizens' organizations and the utility. This time, the room at the Régie's headquarters in Montreal's Tour de la Bourse was packed with at least a hundred participants—including representatives of Russian and Chinese companies—their eyes glued to screens and phones as they listened to a parade of presenters argue for or against letting crypto-miners hook up to Hydro-Québec's grid.

Among the presenters were mayors of several Quebec cities, including some who had already signed agreements with crypto-mining companies to transform industrial properties into server farms. The North Shore city of Baie-Comeau had even started building a new "technological park" to host crypto-mining companies. "We are trying to move away from pulp and paper to build a diversified economy, one that's innovative and responsible," the mayor of Baie-Comeau, Yves Montigny, explained to the Régie's commissioners. Montigny claimed crypto-currency was the only way to end the dependency of his city, with its population of 22,000, on heavy industry. The mayor was livid that Hydro-Québec might put the brakes on his plans, and dismissed the idea that crypto-mining could lead to power shortages across the province. "I was elected with a clear mandate to work on economic development in Baie-Comeau and crypto-currency is part of our plan."

Yet crypto-currency mining is clearly no miracle solution for economic development. Vladimir Plessovskikh, CEO of GPU.one, the very company that was hoping to set up shop in Baie-Comeau, argued that in exchange for five megawatts of power supply, he would be creating 11 new jobs. That's 2.1 jobs per megawatt of power. Among server farms this rate puts GPU.one on the high end of job creation. But a study commissioned by Hydro-Québec by the auditing firm KPMG concluded that overall, crypto-mining created as few as 0.4 jobs per megawatt of power. By comparison, aluminum smelters create 2.2 jobs per megawatt; steel plants, 8.8 jobs and traditional mining, 27 jobs per megawatt. The traditional data centres Hydro-Québec was already working

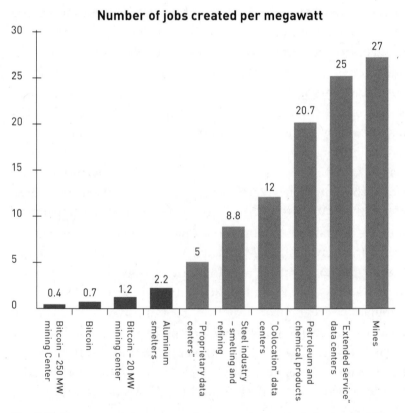

Fig. 8-B: How many jobs are created per megawatt of electricity? The table shows how much more energy crypto-currency mining consumes than aluminum smelters.

to attract would create many more jobs. A centre storing data for its own needs produces 5 jobs per megawatt. Data hosting enterprises create around 12 jobs per megawatt. Centres with wider vocations (including research, call centres and service centres) create double that.

In April of 2019 the Régie de l'énergie decided that the rates for crypto-mining companies would be between 3.46¢ and 5.03¢ per kWh. However, this was accompanied by a set of limitations. The total number of projects could not amount to more than 300 MW and no single project could use more than 50 MW. Hydro-Québec also has to set aside 50 MW for projects of less than 5 MW. Only the most deserving projects would be allowed to hook up based on a list of criteria that included number of jobs created, total payroll, revenues and amounts invested—all broken down by megawatts of power. The companies also had to promise to remain in operation for five years.

Electric Cars on the Right Track

The electrification of transportation is the other avenue Hydro-Québec is banking on to increase energy sales in Quebec. In this case, no one expects Hydro-Québec to be in a situation as urgent as the one the crypto-mining industry created. On the contrary. To make those kinds of waves in Quebec, it would take no less than a million electric cars on the province's roads. And that's not going to happen tomorrow.

Few are aware of it, but Quebec is the uncontested Canadian leader in electric car sales, with almost 40,000 electric cars or rechargeable hybrids on the province's roads, and more than half of the Canadian sales in this category. Granted, electric cars only account for half of 1 percent of the cars sold in Quebec. And the province is still a long way from Norway, where electric cars and rechargeable hybrids account for half of new-car sales. The Quebec government set a goal of having 100,000 electric cars in the province by 2020; 300,000 by 2026 and 1 million by 2030. A million cars would require a total of almost three terawatt-hours of power. "At the moment, 2500 electric cars are sold in Quebec every month," says France Lampron, Director, Transportation Electrification at Hydro-Québec. "So yes, 100,000 cars by the end of 2020 is a realistic goal."

Quebec's electric car capital is Rawdon, population 11,000, a small city in the Lanaudière region about 60 km north east of Montreal. The city has 36-year-old Hugo Jeanson to thank for its title. Jeanson is the co-owner of the family-owned car dealership Bourgeois Chevrolet, founded in 1964. The dealership today sells more electric cars than any other in Canada—almost 400 per year, of all makes—and electric cars account now for half its overall sales figures.

Hugo Jeanson discovered electric cars almost by accident in 2013, when he accepted a Chevrolet Volt as a trade-in from one of his customers. "I was doing them a favour," says Jeanson. But he was curious, and decided to try the car out. He was instantly won over. "I lived closed to Pierrefonds at the time, in Montreal's West Island. So the round trip to Rawdon cost me $100 in gas per week. Driving an electric car cost me almost nothing." Convinced other commuters would be won over to electric cars, Jeanson started accepting more trade-ins of used electric cars, then started training his mechanics and sales people in the field, then started selling new electric vehicles. "People were coming from all over Quebec, Ontario, New Brunswick and even Western Canada to buy electric cars from us. They wanted to do business with a dealership that really understood them," says Jeanson. Today it no longer takes his sales people two hours to explain electric cars to potential customers. "Customers know more about them," says Jeanson. The cars are gaining more autonomy, the technology is improving and the prices are coming down.

If Quebec leads the parade in electric vehicle sales, it's partly because electricity is so cheap: "filling" a battery in Quebec only costs a dollar or two. But like California, the Quebec government offers generous incentives to buy electric cars: an $8000 subsidy for a new electric car under $75,000 and $4000 for a used car, plus $600 to install a residential charging station. The perks for electric car owners are best of all in Laval, where residents get an extra $2000 subsidy when they buy an electric car. Quebec law now requires car dealers to keep an inventory of electric cars, and starting in 2018 will require 3.5 percent of vehicle sales to be in the "low emissions" category. No other place in Canada currently offers anything close to these kinds of incentives.

That said, when it comes to offering perks to electric vehicle owners, Quebec doesn't hold a candle to Norway. In the Scandinavian country—which, curiously, is also a big oil produ-cer—electric vehicles can use lanes reserved for buses and taxis; they have special free parking spots with charging stations, and are exempt from certain road tolls. Norway will even be forbidding the sale of combustion motor vehicles, starting in 2025.[1]

Following in Norway's footsteps, the Quebec government is working hard to remove the principal psychological barrier that prevents wide-scale adoption of electric cars: range anxiety. "It's a well-documented problem, but it has a solution," says France Lampron. Even though more than 90 percent of charging is done at home, even though the average Canadian drives only 41 km per day and even though people can drive 100 km on today's batteries, people are afraid of running out of energy. It's this risk of not hav-ing enough power in exceptional cases that causes anxiety among potential buyers. The market will take off when people are sure they'll be able to find a charging station whenever they need one.

Quebec is moving in that direction. Hydro-Québec began building a network of charging stations in 2011 in partnership with various private companies—beginning with Quebec's hardware chain Rona, grocery store Metro and the Saint-Hubert chain of rotisserie chicken restaurants—and the city's transport agency, the Agence métropolitaine de transport. At the end of 2017, the

"Circuit électrique" (electric circuit) had 1543 charging stations, plus 146 high-speed public battery charging stations. In addition, there are at least as many private charging stations at different companies throughout the province. The Circuit électrique now has 318 partners, including private companies, municipalities, universities and Quebec colleges (CEGEPs). "The fact that private companies joined the Circuit électrique is quite unique," says France Lampron. "In the United States, in Georgia for instance, the electric company offered to *give* companies their own charging stations but the companies wouldn't accept them. Private companies in Quebec, meanwhile, really want to be part of the Circuit électrique."

In 2018, Hydro-Québec launched a Circuit électrique 2.0, which will add fast-charge stations that only take 20 minutes to "fill up" cars, as opposed to 4 to 6 hours at regular charging

Fig. 8-C: "Range anxiety" is one of the primary obstacles to the wide adoption of electric transport. Quebec's government asked Hydro-Québec to build a permanent network of rapid charging stations to encourage the transition to electrified transport.

stations. In Norway, fast-charging stations were key to increasing the number of electric cars on the road. In June 2018, the Quebec government asked Hydro-Québec to take over building the network of fast-charge stations. Hydro-Québec will be spending $10 million a year installing and maintaining 1600 high-speed charging stations over the next 10 years. "The government is asking us to make a permanent commitment to install, maintain and replace the fast-charging stations," says France Lampron. The government even voted a special law with clear instructions to the Régie de l'énergie requiring it to comply with government standards to reduce greenhouse emissions in its decision-making, as opposed to only considering how different factors influence the price of energy, as it normally does. The text of Bill 184 reads, "The Régie shall also consider such economic, social and environmental concerns as have been identified by government decree."

In Éric Martel's view, the reason Hydro-Québec should be in charge of building and maintaining fast-charge stations in Quebec is simple: "Ninety-two percent of battery charging is done at home. Only 8 percent is done at public charging stations. Because of that, public charging stations are of no commercial interest for anyone except a company that sells electricity in the first place. For Hydro-Québec the stations become even more interesting because they remove the principal source of resistance to electrifying transport, which is range anxiety."

"Quebec will be the only jurisdiction in North America where a network of charging stations will be planned out for an entire territory, then maintained after it is built. It's not being done anywhere else. Ontario handed the job to municipalities but there is no coherent, overall effort. The United States is the same story. The only coherent network is Tesla's," explains France Lampron, who has the task of the figuring out how to cover Quebec's territory, possibly with the assistance of partners from the private sector. "The sites that made sense for ordinary charging stations won't necessarily work for the new stations. It doesn't make sense to put fast-charging stations that recharge car batteries in 20 minutes next to movie theatres, for example, because people are going to spend two hours there, anyway. We think it will make more sense to put fast-charging stations near grocery stores, or in gas stations, where people are in a hurry."

Now that the network is being built, Lampron is starting to think about solutions for Hydro-Québec's next challenge, accommodating 100,000 electric cars in 2020, 300,000 in 2026 and up to a million in 2030. These numbers are promising for increased electricity sales, but they will also exacerbate the famous power deficit Hydro-Québec is struggling with, especially if people decide to plug their cars in at the same time, when they get home from work. A million cars charging at the same time, with each using 12,000 watts per car, will add up to 12,000 megawatts of power or 30 percent of Hydro-Québec's power capacity. It's an enormous amount of power. For Hydro-Québec, the trick will be finding a way to avoid overloading the system at specific times—probably by offering a service allowing drivers to charge from a distance using remote control.

"We might also be able to manage the problem with a rate differentiation program," says France Lampron. "But for Hydro-Québec, electric cars offer another possibility: using car batteries as electric reservoirs." In Varennes, Hydro-Québec is studying scenarios that would allow Hydro-Québec to use car batteries to reduce peak demand at critical hours. "A million electric cars would represent a huge load, but together, they could become energy reserves as well."

Chapter Nine

The Battery of the North East

When we stepped out of the elevator on the 18th floor of Hydro-Québec's headquarters, we were surprised to find ourselves in the middle of a trading room. At first sight, it looked exactly like every other trading floor we'd seen: employees talking on phones, eyes fixed on the blinking screens in front of them. But this one is different. The "merchandise" being traded here is not currencies, securities or commodities, but "energy blocks."

Hydro-Québec's 18th floor is the biggest energy-trading floor in Canada. On the spot market, energy prices vary by the minute according to an auction system that works on supply and demand, which are, in turn, determined by weather conditions and what generating facilities are available at a given time and place.[1] The 50 employees on the floor watch energy price fluctuations and weather conditions as far away as the U.S. Midwest, and sell energy as far as Tennessee and Indiana. At any given time of day, they know which generating stations are operating, which ones are under repair, where power lines are working and whether any have been tripped by an overload. It's a cut-throat, volatile market. In a few minutes, prices can climb from 6 to 1000 USD per megawatt-hour. One minute Hydro-Québec is selling 500 megawatts to Toronto for 3 hours at $75 per megawatt-hour; the next, it's selling 1200 mega-watts to New York for 30 minutes at $45 per megawatt-hour.

Fig. 9-A: Hydro-Québec's profits heavily depend on what happens on its energy-trading floor, the biggest one in Canada. The 50 traders and analysts who work there are constantly monitoring weather conditions and the state of neighbouring networks in order to sell megawatts at the highest price.

Since it opened in 2000, the energy-trading floor has been key to Hydro-Québec's financial health. In 2017, Hydro-Québec exported about 34 TWh of power, or 16 percent of energy generated. However, this 16 percent made Hydro-Québec $780 million, a full 30 percent of the company's total profits. Hydro-Québec is a major player in the energy-trading market: New England depends on Quebec for 15 percent of its total electricity consumption; Vermont, up to 25 percent; New Brunswick, 15 percent; New York, up to 5 percent; and Ontario, 4 percent.

Since the 1980s, exports have played an increasingly important role in Hydro-Québec's financial performance, and that role is expected to keep increasing in the next 25 years. Because of the stagnation of Quebec's local electricity market since 2007, exports have become one of Hydro-Québec's best bets for future growth. It was no surprise that shortly after he was elected Premier of Quebec in the fall of 2018, François Legault headed straight to Toronto and Boston to shake the trees.

Yet Quebeckers, for the most part, have a poor understanding of the logistics of electricity exports—at least that's what we concluded after listening to journalists' questions after Éric Martel announced Hydro-Québec had signed a contract to sell 9.45 terawatt-hours (TWh) to Massachusetts for 20 years. At a press conference announcing the deal in January 2018, journalists asked basically the same question half a dozen times in different terms: "If exports are so profitable, why doesn't Hydro-Québec reduce its rates in Quebec?"

The answer takes us back to 1981, when Hydro-Québec's vocation changed. That's when the Quebec government gave the company an explicit commercial mission (see Chapter 3). Since then, Hydro-Québec's profits translate into dividends that are paid to the government, where they go toward the collective well-being of Quebeckers, not reducing electricity rates. Hydro-Québec's CEO summed this up at the press conference: "Export profits are paying the salaries of Quebec's nurses and teachers." And in fact, the dividends are only what is left over after Hydro-Québec has used profits from electricity exports to pay for regional development projects, following orders from the Quebec government. Recent wind power supply contracts are a good example of this dynamic: though they will end up costing Hydro-Québec money, their main goal has been to encourage economic development in the Gaspésie and Bas-Saint-Laurent regions.

Meanwhile, the Quebec press also criticizes Hydro-Québec for energy "dumping" practices, accusing the company of exporting its electricity at "bargain basement prices," $46 per megawatt, while residential Quebec customers pay a higher rate of $59.10 per megawatt (which is the price Hydro-Québec Distribution pays to buy energy from Hydro-Québec Production). The comparison looks bad until one realizes it's faulty. The price neighbouring systems pay is a wholesale price for power that includes production and delivery to the border, but not transmission and distribution. When these costs are added, American consumers end up paying much more than $59.10 per megawatt for electricity. The price is more in the range of $100 or even $150 per megawatt. The *wholesale* price Hydro-Québec Distribution pays (when it buys power from Hydro-Québec Production) is actually in the range of $25 per megawatt. The result: Hydro-Québec Production makes almost 100 percent profit on its exports.

However, while exporting electricity is profitable, profit margins are on the decline. That's mainly due to the explosion of shale gas production in the United States, which dramatically reduced the price of electricity generated using gas. "In addition to causing a drop in the market price for electricity, the shale gas surplus set back the progress of solar and wind power by years," says Ontario energy consultant and electricity specialist Tom Adams, President of Tom Adams Energy.

Indeed, Hydro-Québec's export opportunities are getting harder to find. "If the trading floor continues to make as much profit today, it's only because we are selling twice as much energy as we were 15 years ago," explains Simon Bergevin, Director, Energy-Trading Floor. In May 2016, President and CEO Éric Martel announced a plan to double Hydro-Québec export sales by 2030 by acquiring new infrastructure in other countries and commercializing Hydro-Québec's patents (see Chapters 10 and 11). But exports, which already represent $1.6 billion in sales per year, will be the foundation for this growth. According to projections in Éric Martel's business plan, Hydro-Québec can hope to triple its export sales by 2030. That's right: *triple*. At least. "Our predecessors left an incredible legacy, but we are facing very different challenges today than they were. Since 2007, energy consumption in Quebec hasn't increased. Exports are our best bet for future growth," Martel told us in an interview.

Tripling export sales doesn't necessarily mean tripling the *amount* of electricity Hydro-Québec sells. In the long run, Hydro-Québec sees itself not simply exporting, but playing an active, structuring role in the North American energy market. It would like to become the "Battery of the North East." How, exactly? Hydro-Québec, which is one of the biggest electricity companies on the continent, is already starting to operate like a huge battery. The company's 28 reservoirs act like giant energy storage batteries that can stock energy not just from flowing water, but also from the wind, sun or even other power systems. The weak point of all other forms of renewable energy sources is that: unlike hydroelectricity, they can't be used to store energy, even for a second. The unique property of hydroelectricity (holding potential energy in the form of water in dams) puts Hydro-Québec in a position to become the cornerstone of a "network of networks" that use it as a battery to store their energy.

Becoming the "Battery of the North East" is a new possibility Hydro-Québec is just starting to think seriously about. Meanwhile, it's using its other strengths to develop energy exports. Hydro-Québec uses energy that is totally renewable. The fuel its generators run on is free and will literally keep falling from the sky as long as it rains. Those factors make the price of hydroelectricity foreseeable for the duration of any potential contract—a great selling point when Hydro-Québec reaches out to other electric utilities.

But boosting exports will have consequences for Quebeckers. "If Quebeckers were able to waste less energy, Quebec would be able to export twice as much as it does now," says Pierre-Olivier Pineau, Research Chair in Energy Sector Management at Montreal's HEC business school. This is actually the reason Quebec's government promotes energy efficiency policies so strongly, even if Hydro-Québec has energy surpluses at the moment. Éric Martel estimates that by applying energy efficiency measures Hydro-Québec could easily save 17 terawatt-hours, or 10 percent of present demand. In Pineau's view, the savings could potentially be twice that, or around 30 terawatt-hours. "We could resell this energy outside of Quebec at a huge price, or use it to operate more factories here rather than heat backyard pools in the summer."

So what's holding Quebec back from just opening the valve and exporting more electricity? Well, there's a catch: exporting electricity is not nearly as simple as filling up an oil tanker. It can only be transmitted through interconnected transmission lines. Quebec simply doesn't have enough of them of them right now.

The Interconnection Dilemma

Hydro-Québec benefitted more from energy deregulation in the late 1990s than most other companies. One reason was that it adapted its own structures remarkably well to the new set of rules. Another was that it invested heavily in its energy-trading floor to add personnel and increase its technical capacity. In addition to the electricity traders, the 50-person trading floor team has meteorologists, economists and lawyers who carefully watch the weather and market activities and keep an eye out for changes in regulations that might work to Hydro-Québec's advantage. "Information is key and we look everywhere for it," says Simon Bergevin. "This can go as far as having monitoring services that film steam coming out of thermal generators so traders can see if the generation level is increasing in another system. That, in turn, gives us an idea of how quickly those generators will be able to react to weather fluctuations."

The competition from U.S. shale gas lowered prices on the energy market overall, making it harder for traders to find opportunities to sell. But the biggest factor slowing things down for Hydro-Québec remains its lack of interconnections. Simon Bergevin's colleagues literally pull their hair out every day trying to solve this problem. Sometimes Hydro-Québec's traders have to pass up opportunities to sell when prices are extremely high because there's just no way to deliver the merchandise.

There are presently 15 interconnecting lines crossing Quebec's borders: eight to Ontario, one to New York, three to Vermont and three to New Brunswick (see map, p. 13). Their total capacity is

8000 megawatts, which is 20 percent of Hydro-Québec's overall capacity. While that may sound like a lot of lines, but it's a bare-bones structure: on the borders with New York and New England, which together account for three quarters of Quebec's exports, the interconnections can only transmit *half* of the 8000 megawatts Hydro-Québec has available to export.

Part of the traders' work is finding ways to get through energy "traffic jams" on the available transmission lines. These electricity bottlenecks happen at the border, but also across the border, in U.S. territory, where many systems are struggling with their own transmission-line shortages. In New York, the problem is so serious that extra loops have to be opened as far away as Maine, Indiana or even Wisconsin. Those are big detours.

To solve the problem, Transmission Developers, an American company based in Albany that specializes in energy transmission, is looking to build an underwater cable under Lake Champlain and the Hudson River that will link Hydro-Québec to New York City. According to Gary Sutherland, Director, External Relations at Hydro-Québec International, the $2.2-billion project with the catchy name, "Champlain Hudson Power Express" should have been built long ago. "The last big interconnecting line linking Radisson to Boston was built 30 years ago. That's why the future export contract to Massachusetts is so important. A new transmission line is part of the deal."

In 2016, when Massachusetts announced it would be looking to sign contracts for the large-scale purchase of renewable energy, Hydro-Québec jumped on the opportunity, preparing six versions of its offer. "The interest of these contracts was more than just the value of the energy being sold," explained Sutherland. "From the outset, Massachusetts was offering to pay for a dedicated transmission line. So even if the contract wasn't renewed after it expired, the connection would still be there."

On January 25, 2018, the Governor of Massachusetts surprised everyone by announcing that Hydro-Québec had been chosen to supply the biggest of the energy blocks, 9.45 TWh per year, and that the new line would pass through New Hampshire—an investment of $1.6 billion. By all reckoning, it is an enormous contract, representing almost 15 percent of energy consumed in the State of Massachusetts. Hydro-Québec's directors were pleasantly surprised by its size. Starting in 2020, Hydro-Québec would be exporting $500 million of energy per year for 20 years, at the price of 5.9¢ US per kilowatt-hour. At the exchange rate for the U.S. dollar at the time, that is almost three times Hydro-Québec's cost price.

Then February 1, 2018, there was a complete turn of events: New Hampshire refused to allow the power line to pass through its territory because it would go through its White Mountain National Forest. The energy company Eversource, Hydro-Québec's partner responsible for the construction of the transmission line south of the border, had failed to get the citizens of New Hampshire on board with the project.

"The problem we have in New England is a structural one," explains Gary Sutherland. "The highest consumption is in Boston and Connecticut, but for the power lines to get there, they have to pass through the border states, who aren't big energy consumers. The border states will benefit from the project in other ways. They'll get access to electricity from a renewable source at a reasonable, stable price—something that would improve the carbon footprint of the whole region. But this kind of argument doesn't appeal to people nearly as much as the promise of jobs building windmills does."

In the spring of 2018, Massachusetts announced that the contract with Hydro-Québec would be going ahead anyway. The energy would transit through Maine, a much less controversial route since it would use existing lines, and any new lines that had to be built would go through forestland (where there are fewer landscape issues). Even in favourable conditions such as these, Hydro-Québec will have to negotiate its way through a complex

Fig. 9-B: A group of employees looks tiny standing in the Outaouais substation. Selling electricity to neighbouring systems means Hydro-Québec will require large investments.

regulatory maze that will involve state, counties and towns. The federal government will also have their word.

In spite of all the and downs of the Massachusetts contract, Quebec's profile as a hydroelectricity generator has benefitted from changes in the U.S. market. In fact, the situation is a complete reversal. For the last 30 years, Americans have demonized Quebec's electricity. Reporting on Quebec energy was always highly emotional: when Hydro-Québec proposed its dam project on the Great Whale River in James Bay at the beginning of the 1990s, the company was portrayed as a pitiless monopoly bent on destroying Quebec's indigenous populations. A generation or two ago, this characterization would have held some truth. To this day, the U.S. lobby of gas producers, who support ecological groups like the Sierra Club, never miss an opportunity to trot out old stories. "Americans have always found reasons to claim that any electricity coming from outside the United States is not ecological. The principal reason they do this is that any (new) hydroelectricity produced elsewhere, like in Canada, represents competition for American producers, whether hydroelectric, gas, coal or nuclear," explains Roger Lanoue, former vice-president, research and strategic planning at Hydro-Québec and co-chair of the Quebec Commission on Energy Issues in 2013.

But things have changed. The wind started turning in Hydro-Québec's favour in 2015, leading up to the signature of the Paris Agreement. That year, American economist Jeffrey Sachs, who led a study group on decarbonization, published an influential report called Pathways to Deep Decarbonization. "Their conclusion was that, without outside help, the North East United States would never manage to meet the ambitious decarbonization goals of 80 percent reduction in 20 years," says Gary Sutherland. "Even if they went full speed ahead with solar and wind energy, it wouldn't be enough." Coincidentally, many nuclear generating stations in the U.S. and Ontario are on the verge of being downgraded and decommissioned, leaving governments very few other options.

This new context in the U.S. is a boon for hydroelectricity, which is the only source of renewable energy that satisfies all new environmental, economic and technical criteria. The struggle against climate change has meant all Quebec's neighbouring systems are now trying to reduce their reliance on the hydrocarbons sector, whether its oil, coal or natural gas, and substitute these with energy from the sun, wind or water. As well, although shale gas is cheap and abundant, the case of Massachusetts demonstrates that certain buyers are willing to pay a higher price to secure a supply of renewable energy. Hydro-Québec is already contemplating other long-term contracts.[2]

The New Service of Balancing

The transition toward renewable energies in the United States could make it possible for Hydro-Québec to become much more than just a traditional electricity exporter. Hydro could offer new services like "energy balancing" or energy storage, services that would turn the company into the "Battery of the North East."

Every power system faces the challenge of balancing supply and demand. Demand varies according to weather and consumer needs. Now the massive introduction of solar and wind energy has also made generation (and therefore supply) variable. That makes it even harder to system managers to balance supply and demand. "Two years ago, hardly anyone was talking about the impact of solar and wind on the network," recalls Simon Bergevin. Everything changed when the popularity of these energies started exploding, increasing their effects on power systems. "Now it's part of every discussion."

Hydro-Québec is contemplating offering energy "balancing services" that would allow neighbouring systems to stabilize their grids by compensating generation variations. There's actually nothing new about this idea: "balancing" was part of the ancillary services that were put in place when the energy market was deregulated in the 1990s. Since then, the North American market has been sliced into various "power exchanges" whose objective is to facilitate trading. These exchanges must also ensure there are a certain number of services in place that allow the regulation of the system as a whole. It's a matter of ensuring that the basic parameters (voltage, frequency, quality of the wave) are constant, in addition to being able to maintain a reserve, which can be mobilized to compensate for fluctuations. It's an old problem all power systems face, but it has been complicated by the massive introduction of wind and solar energies, whose generation is less constant.

As Éric Martel explains, Hydro-Québec already offers balancing services to wind energy generators in Québec. "They supply us with energy, but since it is variable, we devote certain resources to helping them balance their output. We could offer the same service to neighbouring systems, to help them make the energy transition. We can't balance every system everywhere, but we could do it for quite a number of them."

In the context of a massive transition toward renewable energy sources, only an entity with a copious source of energy and solid infrastructure can offer balancing services on this scale. "To solve its balancing problems, California, for example is looking to establish links with hydroelectric generators that are very far away, like BC Hydro or other markets in the U.S. Midwest," Sutherland explained. "The same problem will come up in the North East, where the big hydroelectric producer is right across the border."

"When there's no hydroelectric power available, the balancing service for the wind and solar energy has to be provided by gas-fired generation. When you stop and think about it, it sort of defeats the purpose," Éric Martel told us. In the United States, where all hydraulic capacity has been developed, the hydroelectric industry is looking at building double reservoirs to store energy. There will be a horizontal reservoir (an artificial lake), as well as a vertical one, a pumped storage facility. Any excess energy available during periods of low demand will be used to pump water up the vertical reservoirs. The water can then be released to produce extra power when it's needed. Hydro-Québec would never have to go to these lengths to build its water reserve: its extra water is pumped for free by an extremely sophisticated system of evaporation-precipitation known as "rain."

Nevertheless, Hydro-Québec will have to answer a number of questions before it can offer a balancing service. In addition to building interconnections, there is the issue of price. No one knows how much to charge for this kind of service. "We'd also have to convince the managers of neighbouring systems to trust an external power system. They are used to solving their energy storage problems internally," says Sutherland. "On the other hand, when they make the massive shift to alternative energies, we're quite sure they won't have any choice but to look outside their own grid for solutions."

Gary Sutherland is convinced that when it comes to the matter of balancing, Hydro-Québec has a few aces up its sleeve. "Contrary to gas, we're offering a renewable energy source. With hydroelectricity, it's also possible to stock surpluses," he explains. In other words, Hydro-Québec won't necessarily just export electricity: it could also regularly import energy surpluses from neighbouring systems, store them, and then send them back when they are needed. Such an approach solves an old problem in conventional thermal or nuclear energy systems: these types of generators can't be used in emergency situations or when a fast response is required because it takes too long to start them up (and shut them down); for instance, when demand decreases in Ontario, nuclear generation stations can't be shut down quickly. This creates an energy surplus that the province has to get rid of quickly. The surplus can also be increased by a sudden upswing in wind generation, which is totally unpredictable. In the present situation, Ontario Power Generation has no choice but to sell its excess energy at a loss—meaning it actually *pays* to get rid of it.[3] Hydro-Québec takes advantage of the situation in Ontario by putting big hydroelectric generating stations into low gear, or even shutting them down for several hours and letting Ontario feed the Quebec system. All Quebec's neighbouring systems have the same problem: there's a time when they don't know what to do with their extra electricity. If Hydro-Québec operated like a battery, neighbouring systems could, simply, literally, "store" their extra energy in Quebec and get it back later, when they need it. Hydro-Québec could just export or import at will, provided that lines are available for that purpose.

In 2016, Hydro-Québec put this practice on paper for the first time in a contract with Ontario. When the agreement was unveiled, the media zoomed in on the fact that this was the first long-term electricity supply contract Ontario had ever signed with Quebec. It was a relatively small amount of energy, about two terawatt-hours—roughly the equivalent of what Hydro-Québec exports to New Brunswick. However, the contract also included storage and balancing services. For example, Hydro-Québec will receive 500 megawatts of surplus power from Ontario in the winter, which it will store and then return to Ontario in the summer.[4]

There wasn't actually anything revolutionary about the contract. It just put a practice that had been going on between Quebec and Ontario for years on paper for the first time. "People often say the interconnection lines between Ontario and Quebec are underused," says Tom Adams. "But that's because the two systems have always used them in a spirit of mutual assistance, a little like as if each system was an insurance policy for the other."

Fig. 9-C: A convoy of trucks at the U.S. border in September 2017. After Hurricane Irma, 125 Quebec line workers travelled to Georgia to repair damage from the storm. Even if the systems are fiercely independent, they agree to help one another out in emergencies.

"This contract was a little out of the ordinary. We wanted to test the balance and storage concept to see how it would work in the form of a contract, especially given what's coming in Ontario," explains Éric Martel. In the next five years Ontario will have to refurbish no less than half its nuclear reactors. If it turns out that Ontario has to shut down more reactors than expected, like it has in the past, the province will have big energy needs to fill in the short and possibly the long term as well. A lot of Ontarians feel energy rates are spiralling out of control, and they're starting to wonder why they can't take advantage of Quebec's energy capacity—especially since the existing interconnections with Quebec are underused—rather than putting up solar panels at home.[5]

The Integration Dream

The "battery" role Hydro-Québec proposes playing for its neighbours is actually a step toward the integration of power systems in the region. Economists and system managers have been dreaming about this moment. Unfortunately, the idea of integrated markets has never sat well with either politicians or voters for the simple reason that societies tend to be protectionist about electricity.

Still, a number of places have managed to create integrated electricity markets. In 1996, the power systems of Scandinavian countries created a genuine common market for electricity, similar to the Auto Pact, the trade agreement between Canada and the United States that removed tariffs on automobile production. As early as 1927, Pennsylvania and New Jersey integrated their power systems; then Maryland joined in 1956, followed by 11 other jurisdictions including North Carolina, Delaware, the District of Columbia, Illinois, Indiana, Kentucky, Michigan, Ohio, Tennessee, Virginia and West Virginia.[6]

This type of integration essentially creates a "common market" for electricity. The logic behind integration is that individual systems are not managed like autarchies but as part of a whole. The idea is something like a large complex of electric condominiums where everyone owns their own home, but the services are shared. For the integrated systems the issue is not where generating stations and power lines are built; the point is making the system reliable, secure and efficient by eliminating redundant systems and duplications and by increasing interconnections to optimize the use of available resources and equipment.

This kind of thinking is still far from the norm. When it comes to organizing power systems, most American states and all Canadian provinces jealously protect their independence and restrict sharing to a bare minimum. "It's strange when you think about it. Ontario imports gas, oil and uranium, but they won't accept the idea of importing electric energy," say Pierre-Olivier Pineau. "New York has the same attitude. New England is becoming more interested in integration because their electricity is getting so expensive. But high energy costs aren't yet translating into open attitudes. Electricity in Ontario is very expensive, but Ontarians are not interested in Quebec electricity, even if it's a lot cheaper. Even in Quebec, where we would love to export our cheap energy, there is still resistance to the idea of full-market integration."

Although Pineau supports the idea of integrated power systems, overcoming political resistance will be difficult, he says. "Quebeckers would never allow a Vermont wind power generator to supply wind power, even if they could make a better offer than their counterparts in Quebec. It would still have to be Quebec wind generators."

According to Pineau, it will never be possible to fully integrate power systems until there is a complete change of mentality—both in Quebec and among its neighbours. Quebeckers have to accept the idea of making their electricity available to others, but they also have to commit to wasting less of their own, and accept the idea of paying higher rates. "At the moment, Quebeckers are very conscious of the incredible value their hydroelectricity could have *for others*. But there's a price to pay if they want to take advantage of this. The question would lead to very heated debates in Quebec."

In an interview with us, François Legault claimed he was willing to go a long way to make energy integration happen, saying, "I'd like to create an 'energy alliance.' In the case of Ontario and Newfoundland, for example, it could take the form of a joint venture to develop integrated projects in the future." Even before Legault's Coalition Avenir Québec (CAQ) party came to power in October 2018, Éric Martel was taking advantage of every opportunity he had to encourage neighbouring utilities to start thinking "regionally" and to take full advantage of hydroelectric power—from Quebec, but also Newfoundland. "Integration is a good basis for future discussions with Newfoundland about what's going to happen when the Churchill Falls contract expires, but it's also a good basis for discussions with Ontario, New York and the Conference of New England Governors and Eastern Canadian Premiers," says Martel.

Meanwhile, energy integration remains a political hot potato in almost every jurisdiction. It's the reason relations between Quebec and Newfoundland suffered so badly over the Churchill Falls contract.

Politics was also to blame for the collapse of an integration project between Hydro-Québec and New Brunswick Power in 2010. In June 2009, Quebec Premier Jean Charest and New Brunswick Premier Shawn Graham unveiled a unique transaction: Hydro-Québec would acquire New Brunswick Power for $3.2 billion. It seemed like a good deal at the time, since it would both free the maritime province from an energy debt that was spiralling out of control, and ensure a stable price and secure the supply of electricity for the duration of the contract. From Hydro-Québec's perspective, the deal would have increased its clientele by 10 percent and given the company access to the Boston market, not to mention direct access to Nova Scotia, which is heavily dependent on coal.

To top it all off, Quebec would have gotten access to Prince Edward Island, which would have made it easier to reach the province's nearby Magdalen Islands.[7]

Then, nine months later, Hydro-Québec unceremoniously pulled the plug on the project. The official reason was that Hydro-Québec had discovered some pre-existing problems: many of the dams in New Brunswick were in worse shape than expected, which would have required negotiating the price of the contract down. The truth was, the transaction had turned into a political catastrophe: no issue in New Brunswick since the 1960s had ever stirred passions like this one. To make matters worse, Graham had been elected in 2006 on the promise that he would never sell NB Power. The Quebec government didn't do anything to sweeten the pill. For instance, it could have presented what was a takeover as a "merger," instead. But New Brunswick was a few months from elections and the opposition was swearing it would renationalize NB Power should the deal come through. In the circumstances, citing unforeseen costs seemed like an elegant way out for everyone.

There may be no use crying over spilled milk, yet the case of New Brunswick shows how electricity has a kind of symbolic value that is unmatched by any other energy form. That's why Premier Legault is talking about eventual contracts to supply Quebec electricity to Ontario not as "integration," but as a "joint venture."

Legault will also have to put his kid gloves on to discuss the future of Churchill Falls, or any other joint project with Labrador for that matter. In November 2018, a month after the Supreme Court rejected Newfoundland's last argument against the Churchill Falls contract, Legault invited Newfoundland Premier Dwight Ball to a private dinner to start talking about the future.[8] "The most promising hydroelectric project in the entire North East is in Gull Island, Labrador," Legault told us. Back in 1998, former premiers Lucien Bouchard and Brian Tobin announced a 2200- megawatt project in Gull Island, but it was shelved in 2000, when Quebec's

Fig 9-D: Premiers Jean Charest and Shawn Graham announcing the project to sell New Brunswick Power to Hydro-Québec in 2009. The project was cancelled for both technical and political reasons.

Innu community protested it—no one had thought to consult them first. "Quebec and Newfoundland could become the Battery of the North East," says Legault. "Together."

Chapter Ten

Of Motors and Batteries

There's a little bit of Karim Zaghib in every iPhone, laptop computer and Tesla car. Zaghib, who started working at Hydro-Québec in 1995 after spending several years in Japan with Sony, is one of the most widely quoted scientists in the world with over 85 patents in his name. His inventions are used in almost all lithium batteries being produced.

When we sat down with Zaghib at his office on the IREQ campus on Montreal's South Shore in Varennes, he broke the ice by scribbling down chemical formulas of different types of lithium batteries for us on a sheet of paper: $LiFePo4$, $LiNiCoAlO2$, $LiNiCoMnO2$ and others. "A battery is an anode, a cathode and an electrolyte, but above all, it's a recipe that requires the right combination of ingredients," he explained. "Certain batteries produce a lot of energy. Others produce more power, or they charge, or run down quickly. Others can be recharged thousands of times. The key is understanding what purpose they serve," says Zaghib, who is General Director of the CE-ETSÉ, the Centre of Excellence in Transportation Electrification and Energy Storage.

Zaghib is extremely proud of the CE-ETSÉ, whose new headquarters opened in March 2018. There are 80 employees working there today under a ceiling covered in black hexagonal panels. "It's the same shape as lithium crystals, which are our raw material," he told us.

Zaghib is not a tall man, 1.65 m at the most, but he has solid shoulders: he is carrying one of Hydro-Québec's biggest hopes for future expansion on them. The company aims to double its revenues by 2030, mostly through exports, infrastructure acquisitions and by commercializing inventions. As we have seen, Hydro-Québec has to look for new ways to expand since its revenues have barely increased over the last decade mostly due to worldwide stagnation in energy demand. If Hydro-Québec wants to grow without hiking domestic rates, it will have to find new sources of revenue, including commercializing its patents.

Fig. 10-A Karim Zaghib, General Manager of the CE-ETSÉ, on the IREQ campus. A renowned battery specialist, Zaghib is one of the most quoted scientists in the world. Hydro-Québec has been interested in energy storage since IREQ opened in 1970.

Of the roughly thousand patents Hydro-Québec now owns, those related to batteries and electric motors have the best commercial potential. Hydro-Québec's researchers have spent the last 40 years working on these technologies. But with the electrification of transportation now at our doorsteps, there is huge potential for Hydro-Québec to tap into a surging market. The rise of alternative energies is making it all the more urgent to develop electricity storage systems—meaning, of course, batteries. "For battery sales alone, we estimate the world market will be in the range of $225 billion in 10 years. If Hydro-Québec gets even 1 percent of that business, imagine what that will mean," Zaghib told us.

Commercializing inventions is an old dream of Hydro-Québec's. CEOs have been talking about their incredible potential for 30 years now. But so far, the promise has repeatedly turned into disappointments. Hydro-Québec is not the only company that has driven down this highway of broken dreams: when it comes to electric motors and batteries, there have been countless "phenomenal breakthroughs" that turned into dead ends. In the 1990s and 2000s, Jean-Benoît frequently reported on these breakthroughs in a half a dozen magazine articles on "promising developments" in the field of electric vehicles. None of them came to fruition.

As 2020 is approaching, however, there's room for cautious optimism. One reason is that the electrification of transportation is taking off for real now. Zaghib's team at the CE-ETSÉ has established itself as a credible player in the field. Sony has sold several hundred thousand LFP storage batteries based on technology licensed by Hydro-Québec. The CE-ETSÉ has been involved in important recent advances in the field of batteries and, when it comes to many new types of batteries, has the advantage of a technical head start.

It's a similar situation at TM4, a partially owned subsidiary of Hydro-Québec that develops electric motors. TM4, of which Hydro-Québec ceded control to the U.S. automotive supplier Dana in 2018, now sells 5000 electric automobile motors per year. At the moment, 5 percent of electric buses in China are running on systems designed by TM4—and China represents 90 percent of the world market for electric buses.

In short, when Éric Martel claims Hydro-Québec could add several billion dollars to its revenues by 2030 by commercializing patents, he's being more realistic than his predecessors since the 1980s.

The main reason Martel's optimism seems well warranted is the change of attitude at Hydro-Québec itself. The story of commercializing patents at Hydro-Québec has been a long saga of dithering in which the government-owned company has not exactly come out a hero. Hydro-Québec's executives and directors have been known to vacillate between periods of hyperactivity and lethargy, sometimes from one month to the next.

That said, transforming a research project into a commercial success is a huge undertaking. Millions of entrepreneurs have failed even after investing mountains of money in their projects. That's because good research is no guarantees of findings. For example, CE-ETSÉ's laboratories are housed in the former headquarters of the "Tokamak" nuclear fusion reactor project (Tokamak is an acronym of the Russian, "toroidal chamber with axial magnetic field"). The Tokamak project was an attempt to recreate in a laboratory the process by which the sun produces energy, in other words, nuclear fusion. Launched in 1986, this fundamental research project, the most important in the history of Quebec science, cost the federal government and Hydro-Québec $130 million. The project failed to meet expectations and the plug was pulled in 1999. The only thing that remains of the project is a set of enormous armoured doors and epoxy-varnished concrete floors where the reactor was housed. "We had to preserve this," Zaghib told us during our visit. "It's part of our institutional memory."

But since 2015, thinking at Hydro-Québec has changed. The arrival of new President and CEO Éric Martel is the main reason.

Robert Baril, President of TM4, noticed the new spirit on his new boss's first visit: "I was making my presentation and Éric Martel asked me out of the blue, What are your sales arguments? What's your sales cycle? What sets you apart?" I could see right away that he understood what we were doing. It showed that he had worked in industry, at Bombardier."

Another former Bombardier executive, Guillaume Hayet, Senior Director, Commercial Projects at Hydro-Québec, now has the task of creating million-dollar companies in the hopes that they will some day become billion-dollar companies. His job is to identify the most promising projects and figure out the best way to bring them to the production phase and get them onto the market. "If Hydro-Québec, instead of Sony, had commercialized the LFP battery, we'd be rich. With the new batteries we're developing, our strategy is to establish ourselves as leaders in the field."

Reinventing the Wheel

Nothing illustrates the dreams and disappointments of commercializing inventions better than the saga of Hydro-Québec's electric motor. The story began in 1980 when Pierre Couture, a researcher at IREQ, revived the work of Ferdinand Porsche. In the 1890s, the automotive engineer who founded the Porsche car company came up with the idea of an in-wheel motor. It was essentially an electrical motor built into the hub of the wheel that recharged when the brakes were used. The concept was revived and applied to the lunar Jeep used in the Apollo 15 mission in 1971. Throughout the rest of the 1980s, Couture worked on the motor and the electronic controls for the wheel. In 1994, he equipped a Chrysler Intrepid with four motor wheels, striking a chord among media and Quebec politicians. Impressed by Couture's work, Hydro-Québec launched an ambitious program to develop the electric wheel and revolutionize the auto industry, only to fold it the following year. "It was one generation too early," says Robert

Baril. "Hydro-Québec understood that for the in-wheel motor to work, the entire direction of the automotive industry would have to change first. In the meantime, battery technology was still a long way from being up to scratch. Developing the wheel would have required billions of dollars in investments."

In 1998, Hydro-Québec created the subsidiary TM4 to re-launch the electric motor project, using a different approach. In 2000, the company decided to develop a central electric motor, which would be much more economical than installing two or four in-wheel motors in a single vehicle. TM4 succeeded in carving out a place among a half-dozen pilot projects of different car manufacturers, but none of them came to fruition. In 2003, TM4 announced it was about to deliver 10,000 motors to the car manufacturer Peugeot, but that project never materialized. Then in 2009, it almost happened. TM4 joined forces with Indian auto manufacturer Tata Motors who were building the Motiv. Yet that project also died.

According to Robert Baril, TM4's real problem was the fact that the company was fixated on the electric car. Sure, electric motors have a great advantage over combustion engines from many perspectives. But the batteries themselves weren't yet up to the task. As far as motors were concerned, no one yet knew how to build one for a comparable price. There was also no after-sales service available, and no market for spare parts. Everything had to be built from scratch.

Hydro-Québec kept absorbing TM4's losses year after year so relations between TM4 and Hydro-Québec remained tense. "Hydro-Québec is an electricity company. It's a government monopoly that never had to fight for customers. Commercialization took it out of its comfort zone," explains Robert Baril, an accountant by training who held different administrative positions at Hydro-Québec before becoming Vice-President of Finance at TM4 in 2009. "Hydro-Québec is not comfortable in the entrepreneurial framework. The questions are different, like who,

Fig. 10-B: A technician at TM4 in Boucherville, Quebec, working on a stator, one of the essential components of the SUMO motor used today in some 100,000 buses in China. Hydro-Québec has high hopes for commercializing its patents for electric motors.

exactly, is TM4's customer? Who decides, the consumer or big automakers? What's the right price? We have to network, reach out to all different players. Our work is not just about innovating and building. We also have to do marketing and provide after-sales service."

In 2011, TM4 was asked to work on the Novabus electric bus project. This was an epiphany. Researchers suddenly realized that electric motors would be perfect for buses and trucks. Since electric motors eliminate the need for extremely heavy transmission and cooling systems, as well as gas tanks (which are, proportionately, much bigger in large vehicles than they are in cars), the weight of the batteries would not pose the same problem in large vehicles as it does in cars. What's more, buses and trucks are managed in fleets. That means that maintenance and charging have to be planned anyway, eliminating many of the unknowns that were stalling wide-scale adoption of the electric car (like range anxiety and fear of not finding parts).

So TM4 created a joint venture with the Chinese manufacturer of alternators Prestolite, hoping to cash in on the growing market for electric buses in China. Since then, the company has developed some 15 models of motors for different types of buses as well as for delivery trucks, mining machines and boats. "Our bus motors have, cumulatively, run 300 million kilometres," Baril told us during a guided tour of the factory laboratory in Boucherville, Quebec, which serves as a research centre for new motors. Since 2009, the 130 employees at TM4, which owns some 60 patents, have managed to reduce the time it takes to get a new prototype ready for testing (for resistance to vibration, temperature extremes, electromagnetic fields) from 24 months to 6 months.

In 2018, with $60 million in sales, TM4 was finally on the verge of making a profit. "But I'm not satisfied," Baril told us in February of that year. "Getting 5 percent of the Chinese market is okay, but I want 15 percent. We have a plan to make that happen." Six months earlier, Baril had crossed the Rubicon by deciding to pull out of the business of car motors, after going nowhere for 19 years. The Hydro-Québec subsidiary had guzzled close to $500 million over 20 years just to get into the market and start breaking even. While that may seem like a long time, it's not actually unusual. Elon Musk's Telsa is the best performing most popular electric car on the market but it is still a money loser. In 2017, Tesla's shareholders had absorbed an accumulated loss of USD 2.2 billion over nine years.

In June 2018, Hydro-Québec announced that it would be selling 55 percent of TM4 shares to the U.S. motor manufacturer Dana Incorporated for $260 million.[1] Based in Maumee, Ohio, the 30,000-employee company was looking for a way to diversify its offer of motors to large auto producers like General Motors and Toyota. The sale price pushed TM4's value to $500 million. Hydro-Québec can consider itself lucky: between 1996 and 2006, the company absorbed hundreds of millions of dollars of pure loss when Avestor, a subsidiary created to commercialize lithium-metal-polymer (LMP) batteries, went bankrupt in 2006. The LMP battery technology was resold to French multinational Bolloré but Hydro-Québec didn't even recover a cent from this operation. "We learned from our mistakes," explains Guillaume Hayet, Senior Director, Commercial Projects. "Hydro-Québec is not the right company to commercialize or manufacture electric motors. Other companies do it better, faster and cheaper."

A number of commentators in the Quebec press criticized the fact that TM4 had passed into foreign hands, but it was good news. The factory laboratory in Boucherville would still be Dana's electric development centre, and for the first time since the beginning of the automobile adventure, TM4 acquired not just better production capacity, but a sales force and a serious visiting card to help it break into the world of big automakers.

According to Éric Martel, Hydro-Québec's goal it to sell 1 or 2 million motors per year, not just 6000. "Producing a million motors is not even the same business as producing 6000. When Hydro-Québec went knocking on doors at GM and Ford to sell its motors, their first question was always "How are you going to support sales and deliver parts throughout the world?" They could see perfectly well that we didn't have the answers to those questions. Dana, however, can make convincing arguments. The marriage with Dana is not just a good deal, it allows us to think big."

The Rise of Batteries

Karim Zaghib's approach to batteries has always been radically different from that of both TM4 and Avestor. First, he has never thought batteries were just about electric cars. He knows they serve other purposes, like storing energy, either in houses or for entire power systems. So Zaghib pursued research on all types of batteries. But Zaghib's vision of battery research was also shaped by the fact that he spent 11 years working on his own, without a penny from his bosses to finance his research. It was not a situation he chose but it did allow him a high degree autonomy, which in the end, paid off.

In 1999, Karim Zaghib had been developing batteries for IREQ for four years when Hydro-Québec declared that its researchers would henceforth concentrate exclusively on its core mission: supplying electricity. Hydro-Québec just scrapped any project that wasn't related to generation, transmission or distribution of electricity. "I am an electrochemical engineer. I didn't see myself working in transmission or distribution. I don't know anything about them," explained Zaghib, who decided to take his special case to Hydro-Québec's CEO André Caillé and to the Director of IREQ. "I made them understand that it was important to be visionary. It was obvious to me that batteries would be useful not just in cars, but in houses and for energy storage. That was already the vision Lionel Boulet, the founder of IREQ, had in 1967. And then, since we were just recovering from the ice storm of 1998, people were starting to think about energy storage as way to make houses autonomous, or even to replace generators. So Hydro-Québec's managers told me, 'Okay, go ahead, but you'll have to get financing elsewhere.'"

For 11 years, Zaghib pursed what he called his "crossing of the desert." He carried on his research with his salary but no other support from Hydro-Québec. Since Zaghib was already a reputed researcher in the field of electrochemistry, he managed to cobble together research budgets from the likes of the U.S. Department of Energy, the European Union and Sony—and never Hydro-Québec for money.

Zaghib had his first real breakthrough in 1999, in the field of lithium-iron-phosphate (LFP) batteries. Sony was immediately interested, and bought a first licence in 2003. The Japanese saw its potential for storing electricity in houses, baseball stadiums and computer and telecommunications centres. In 2009, after developing 26 patents with IREQ, Sony started mass production and IREQ started collecting royalties.[2]

Zaghib came out of the shadows in 2010 when he organized the annual conference of the International Meeting on Lithium Batteries (IMLB) in Montreal, with 1267 participants. The event got Hydro-Québec's attention, and it realized that IREQ had signed licences totalling $26 million for Karim Zaghib's patents. "We suddenly had a prominent spot in Hydro-Québec's annual report," said Zaghib.

In 2018, Hydro-Québec took another step toward commercializing its batteries by creating another subsidiary, the CE-ETSÉ. Even though it's housed on the IREQ campus, the CE-ETSÉ is totally independent. "IREQ's vocation is to do research related to Hydro-Québec's core business. The CE-ETSÉ takes care of what is outside of Hydro-Québec," explains Zaghib, who is reassured, since he's thinking of retiring in the next six or seven years. Work in the areas of electrification and energy storage are less at risk of being compromised by dillydallying on the part of Hydro-Québec's Board of Directors or IREQ's management, which has been the case for other development projects.

Guillaume Hayet, meanwhile, is hard at work building a financial structure, recruiting investors and finding partners for manufacturing and marketing batteries. Hydro-Québec's Board of Directors, having learned the hard way from the TM4 and Avestor experiences, won't be jumping into new business without having solid partners first. "From now on, we will be able to say to investors and to our board, "We will put together financing that will bring a 25, 30 or 40 percent return," explains Hayet. "The thing we won't do is build a factory. It's not in our scope of activity. What we have is the most advanced laboratory in the world and the best team."

Karim Zaghib, meanwhile, is putting together research budgets in the tens of millions of dollars by multiplying collaborations with other institutions. The Lawrence Berkeley National Laboratory, in

Fig. 10-C: Large batteries, the size of a tractor trailer, which contain 576 batteries each. These accumulators designed by Hydro-Québec can supply electricity to 50 houses for 12 hours, or replace a transformer.

California, agreed to give him privileged access to its materials catalogue. The CE-ETSÉ developed nanotechnologies for a new type of battery known as a solid-state battery in cooperation with the NanoBio Lab in Singapore. With Nouveau Monde Graphite, a company exploiting a mine in Saint-Michel-des-Saints, in Quebec's Lanaudière region, the CE-ESTÉ will be developing a process to manufacture made-in-Quebec graphite anodes. The CE-ETSÉ also invented a new five-volt lithium battery with the U.S. Army Research Laboratory in Maryland.

Even if Zaghib no longer works "in the desert," his team still has to chase a lot of rabbits in the next years. It's extremely hard to predict which type of battery will be adopted by which manufacturer, and put toward what use. The number of electric cars on the road is almost sure to grow, but no one really knows when it will take off. Hayet, for that matter, has started a new spin-off, Stockage d'énergie HQ, to commercialize another one of Zaghib's team's "babies": mega-batteries that can supply energy to 50 houses at once. The team is now experimenting with different prototypes: in Quaqtaq, they are testing "small," 10-foot batteries (see Chapter 6), while in a substation in Hemmingford, they have replaced a transformer with two mega-batteries the size of tractor trailers. "We're watching the electric car market closely because it's about to take off. But there are other big opportunities in building, housing and industry," says Zaghib, whose team also invented a Lithium-ion battery that self-charges using ambient light.

"Patience is an important virtue in the business world. It doesn't mean you have to be slow, just that things have to be done the right way, even if it takes six months, or a year, or more," says Éric Martel. "That's the way we think about batteries. We have a very good product but we have to work on reducing costs. And if we manage that, investors will be willing to pay a premium for it."

"We are not talking about becoming the next Kodak here," says Zaghib, referring to the U.S. manufacturer of photography equipment who invented digital photos in 1975, then fatally, fell behind Japanese competitors and declared bankruptcy in 2012. "We have to constantly be on the lookout. The danger is missing out, like Kodak did. They aren't there because they stopped listening."

Chapter Eleven

Hydro World

What would happen if, say, Hydro-Québec bought a power system in the United States, or a dam in Mexico, or a series of transmission lines in Europe? There's no shortage of such opportunities. Governments all over the planet are having trouble building, maintaining and operating their own power lines or systems. They are looking for someone else to take care of the problem, someone with experience, who can do a better job. This is the situation that creates an international electricity infrastructure market.

The idea isn't as far-fetched as it sounds. For instance, a full quarter of the 70 billion euros in sales of France's electric utility Electricité de France (EDF) comes from its activities in 24 other countries: mainly the United States, Italy and Poland and the United Kingdom (EDF controls a large chunk the power generation capacity of Britain). Between 2017 and 2019, Ontario's Hydro One tried (though unsuccessfully) to make a merger agreement with Avista, an energy company based in the State of Washington that also services Oregon, Montana, Idaho and Alaska—a proposed $6.7-billion transaction. In Quebec, Gaz Métro (now called Energir) has owned Green Mountain Power in Vermont since 2007 and acquired Standard Solar, in Maryland, in 2017.

Hydro-Québec International (HQI) actually played this game for a decade during the 1990s—and it played it well. In 2000, HQI's prospects were so promising it looked like the sun would never set on the company's empire. Hydro-Québec owned two hydroelectric generating stations in Panama and in Costa Rica and held interests in 15 other companies in Asia, mostly in China. Hydro-Québec also acquired 660 km of transmission lines in Peru; 180 km in Australia; and 40 km of undersea transmission lines joining Connecticut and Long Island. And the pearl was the 8300 km of lines Hydro-Québec acquired in Chile's Transelec sytem. But in 2006, the sun set on this empire when Hydro-Québec sold the lines at a profit and later returned to its traditional market.

President and CEO Éric Martel took Hydro-Québec International out of storage in 2016. When the former president of Bombardier Business Aircraft became CEO of Hydro-Québec in 2015, he found it a little strange that such a big business with such a good reputation had no interests outside its own territory. Martel then announced his intention to see Hydro-Québec acquire foreign infrastructure. This project was consistent with the expansion strategy he had set out. "I thought we could easily buy systems that were in bad shape, and weren't too expensive. We'd improve them, operate them, make money and it would bring us a new revenue stream. We know how to do that," Martel explained to us in our interview.

Few Quebeckers are conscious of it, but Hydro-Québec is highly respected in international energy circles. Ever since Hydro-Québec left the international sphere in 2006, the World Bank has been regularly inviting it to return to the business, notably to help develop hydroelectricity in a world where only 15 percent of hydroelectric potential is presently exploited. During the United Nations Climate Change Conference in 2015, Martel made good use of his 10 days in Paris to meet the heads of power systems from France, Italy, Norway, China and the United States. Martel took advantage of the opportunity to send the message that HQI was back and was buying. The news spread like wildfire. Since the conference Hydro-Québec has received offers to purchase or build infrastructure every week, whether systems in Latin American, generating stations in Asia or Africa or high-voltage transmission lines in Europe.

In Michel Clair's opinion, it was about time. Former Deputy Minister of Energy in Jacques Parizeau's government from 1994 to 1997, Clair was CEO of HQI from 1997 to 2000. "The internationalization of Hydro-Québec's expertise was, and should continue to be, a core strategy," he says. Indeed, in addition to its technical and financial expertise, Hydro-Québec has its own dedicated research centre whose researchers can find solutions to any new problems that may pop up. Hydro-Québec is the only power system on the continent with its own private research institute, so it can find ways to boost productivity that other system operators can't imagine. "Hydro-Québec has a unique set of advantages," says Clair.

In 2016, Hydro-Québec was on the brink of acquiring a large part of France's Réseau de Transport de l'Électricité (RTÉ), a gigantic transmission system operator with 105,000 km of high-voltage lines (triple the number in Quebec) and an estimated value of almost EUR 8.2 billion. Électricité de France was aiming to sell 49.9 percent of its shares in equal parts to the France's Caisse de dépôts et consignation (the Deposits and Consignment Funds, a public financial institution) and a private investor. Then in July 2016, the French government backtracked, deciding that RTÉ would remain French. As it turned out, the French insurance company CNP Assurances snapped up RTÉ as a "consolation prize," Hydro-Québec and RTÉ agreed to study joint acquisition projects for electric transmission in Europe.[1]

The half-dozen employees who shop for infrastructure for Hydro-Québec International are working on roughly 30 cases at the moment. "There are a lot of opportunities out there," says Gary Sutherland, Director, External Relations at Hydro-Québec International.

Modernización de 60 hidroeléctricas para generación de energía limpia

AFFICHER LA REDIFFUSION DU CHAT

210 812 vues   👍 17 K   👎 128   ↗ PARTAGER   ⊟↓ ENREGISTRER   ...

Fig. 11-A: On December 19, 2018, on YouTube, Mexican President Andrés Manuel López Obrador, pictured, announced that he was entrusting Hydro-Québec with the task of renovating 60 of Mexico's hydroelectric generating stations. If the project goes forward, there will be a lot of people taking Spanish lessons at Hydro-Québec's headquarters. But it can take months, or even years to negotiate this kind of partnership.

Another large-scale project in HQI's second life—and one of the most likely to take off—is in Mexico. Since 2016, Hydro-Québec has been carrying on open discussions with Mexico's government, and more specifically with the Mexico's Comisión Federal de Electricidad (CFE, the government-owned electric utility). In 2013, Mexico's then-President Enrique Peña Nieto launched reforms in the energy sector that included investments of USD 137 billion.[2] The biggest investments were slated for the oil sector, but the reforms also applied to the electric sector. Discussions like the ones between Mexico and Hydro-Québec always have a high degree of diplomacy: Mexicans are very nationalistic, but they know the limits of their power system and understand how it is stunting economic development. The country of 110 million produces only

60 TWh of electricity a year (a third of Quebec's production). Reform of the power system led to the birth of six new electricity generators, but it is still facing serious problems related to capacity, stability and interconnections—Hydro-Québec's areas of strength. Mexico operates many large hydroelectric generating stations; together these account for 8 to 10 percent of the country's generating capacity, but the stations are in poor condition. Mexico's new president, Andrés Manuel López Obrador (nicknamed AMLO) has sworn he will renovate them. It's another potential opportunity for Hydro-Québec. The difference is that now Hydro-Québec is not the only company looking to get into the international hydroelectric infrastructure market. "The Spanish, the Russians, the Chinese, everyone is interested," says Gary Sutherland, who has family in Mexico and is watching the situation closely.

In December 2018, AMLO released a YouTube video of his meeting with representatives from Hydro-Québec, who were accompanied by the Canada's Ambassador and Quebec's Delegate General. In the video, Mexico's President states that he will be doing business with Quebec to repair and rebuild Mexico's 60 hydroelectric generating stations. In the process, Mexicans hope to double their capacity from 13,000 to 26,000 megawatts. But Hydro-Québec knows it's too soon to start patting itself on the back. Competition in the sector is fierce and everything must be carefully negotiated: price, implementation details, the formula for the partnership, supply contracts—everything.

For this type of project, as in any foreign acquisition deal, negotiations are bound to be long and intense. Hydro-Québec established strict selection criteria for foreign projects that go above and beyond getting the right price. Contracts must conform to the company's principles of sustainable development; projects must involve either generation, transmission or distribution of electricity; Hydro-Québec must be able to optimize the equipment it's working on in a way that will increase profits; the countries it operates in must have stable political and legal systems; and finally, the countries must also have ethical guidelines in place for allocating public contracts. "I wouldn't go as far as saying our Board of Directors is allergic to risk. 'Prudent' is the word I would choose," says Sutherland.

The kind of projects that meet Hydro-Québec's strict selection criteria don't exactly grow on trees. In 2017, the Montreal daily *La Presse+* reported that Hydro-Québec had said no to a dam project in Peru because there was too much sediment in the dam's reservoir, a factor that threatened the long-term profitability of the project. Another deal with Brazil's power system fell through because of ethical concerns.[3] "Sometimes it's just one small matter in a long list of issues, but it can turn out to be a problem we don't know how to solve," says Éric Martel. "In Peru, the problem was simply that the dam didn't have a drain plug [a channel], which meant there was no way to get rid of sediment that threatened to clog the equipment."

Two years after stating its intention to venture into the international infrastructure market, HQI still hasn't announced a single project. In 2016, President and CEO Éric Martel sounded like he was being a bit cautious when he announced Hydro-Québec International would invest $100 million by 2020, and $3 to 5 billion by 2030 in foreign projects. As it turns out, 16 years is not much time to get up to speed when you are starting from zero. "It's true that I set targets in 2016, but we're not in a rush and we will proceed cautiously," says Martel. "It will take the time it takes. In any case, our export sector is doing very well at the moment."

Rebuilding from Scratch

"If Hydro-Québec had held on to its international division instead of dismantling it in 2006, Hydro-Québec International would have already met Éric Martel's targets for 2030. It might even have become bigger than Hydro-Québec itself by now," says Michel Clair. "We lost a whole generation, maybe more."

The best adjective to describe the first 28 years of HQI, from 1978 to 2006, would be "erratic." The company explored and tried just about everything, from consultation and development to construction and acquisitions. Its business model was the subject of frequent disputes between HQI, Hydro-Québec's management and the Quebec government. HQI's CEOs lasted, on average, 2 years.

"When I arrived at Hydro-Québec International in 1997, there were only 15 employees, but its engineering consulting business was already active in 75 countries so it had an absolutely incredible network of contacts," recalls Clair. "They were really well connected in ministries and large companies all over the world. It was a formidable intangible asset."

Michel Clair took up his position in 1997 with the goal of achieving the mission set out by Hydro-Québec's Board of Directors at the time: make $3.5 billion in acquisitions in five years. Three years later, in 2000, Hydro-Québec International was on the right path, with a dozen projects in the bag. "It was very ambitious, but we had a strategy, actually many strategies. We knew the Chinese market was going to open up and we got in with an investor from Hong Kong. In South America, our idea was to create a large interconnected market between Chile, Peru, Equator, Colombia and Panama. We had a similar strategy for Central America," says Clair.

Then in 2000, Hydro-Québec's Board of Directors started to waver. The objectives set out three years earlier suddenly seemed too ambitious. HQI's staff of 100 was reduced to 10 employees and Hydro-Québec's three business units, generation, transmission and distribution, were instructed to develop international activities in their own spheres. That strategy produced exactly zero results. "We didn't lack ambition, but we did lack collective maturity," remarks Clair. "Before we jumped into this huge project, no one carefully measured the risks and opportunities, nor the time it would take to create value. People had the mindset of short-term gains."

Fig. 11-B: In its first life, Hydro-Québec International owned a dozen foreign facilities, including Chile's Transelec system (photo). Between 2003 and 2006 it sold everything off. Since 2015, Hydro-Québec has been looking for a way back into foreign markets.

In 2006, after going nowhere for five years, then-CEO Thierry Vandal announced he was shuttering HQI. The subsidiary sold off its last assets in a few months, including the prize jewel, Chile's Transelec network. Together the transactions yielded $939 million in profits, which the Quebec government used as a first contribution to a trust it created to reduce Quebec's public debt, the Fonds des generations (Generations Fund).

No one can undo the past. But there's no doubt that if Hydro-Québec had stayed the course in the international electricity infrastructure market, Hydro-Québec International would already be a major player today. Instead, HQI has to start over in a highly competitive market.

In 2018, Éric Martel admitted he was surprised at the level of competition for international projects. Energy infrastructure is considered one of the safest and most profitable investments out there, so investors from across the planet are ready to fight for a piece of the pie. For every infrastructure project that comes up there are 10, 20 or even 50 buyers. Many are speculators looking for quick purchases they can quickly flip for a profit.

"We're interested in buying for different reasons," explains Gary Sutherland. "Hydro-Québec is a long-term business. We want to acquire assets that will increase in value, but we want to operate them for a long time. With these conditions, the number of possibilities for us is fewer than for other investors."

The fact that Hydro-Québec is already so profitable is another reason the company is selective about future investments. Year in, year out, Hydro-Québec produces profits in the order of 21 percent on sales of $13.4 billion. Very few public companies the size of Hydro-Québec manage to consistently turn profits like that. Hydro-Québec's return on capital, 13 percent, is also very good for a government-owned enterprise. But that means Hydro-Québec has to get its hands on assets that will deliver 12, 15, 20, or 25 percent operating profits or return on capital. Hydro-Québec needs assets that won't tarnish its balance sheet. The question is, where exactly are such fabulous investments to be found?

Ontario's Hydro One has the opposite problem. Its balance sheet does not sparkle. The objective of the $6.7-billion deal to acquire Avista, in Washington State, was to improve Hydro One's overall results. Hydro-Québec would probably not even have considered Avista. Its profitability would have passed muster but the fact that 35 percent of its energy comes from natural gas and 10 percent from coal would have raised eyebrows. Hydro-Québec would have swallowed Avista at the risk of diluting its environmental performance and tarnishing its green reputation.

The acquisitions issue brings another one of Hydro-Québec's challenges to the forefront: protectionism and politicization of energy projects. Protectionism comes into play in all spheres of the economy, but when it comes to electricity, the degree is almost comical (see Chapter 9). Technical-financial expertise is not enough to fight protectionism: it requires refined diplomatic skills, tact and a talent for lobbying. And sometimes these aren't even enough to get results: when Hydro One tried to acquire Avista, the government of the State of Washington just said no, and Ontario had no choice but to throw in the towel. Whether it's acquiring France's Réseau de Transport de l'Électricité or renovating Mexico's electricity network, politics is everywhere in the energy business. "We knew it was going to take a long time to make these projects work," says Gary Sutherland. "And we will take the time it takes."

Chapter Twelve

The Reactive Customer

When Hydro makes headlines, it's generally because someone, somewhere is unhappy—usually either because of a power failure, a larger than expected electricity bill or bad service. Whatever the cause, there's a good chance the phones will be ringing at television stations, at an MP's riding office, or even in the Premier's Office.

It's safe to say that most Quebeckers are inclined to believe the worst of Hydro-Québec. Nothing demonstrated this better than the case of the so-called "smart meters." The story began in 2012, when Hydro-Québec announced it would be doing away with traditional analogue electric meters. From a business perspective, it was a good move: Hydro-Québec would no longer need to send agents to ring on customers' doorbells and maneuver through their basements to get a reading of the number of kilowatt-hours of electricity consumed. The new meters, whose technical term is "communicating meter," would simply send the information back to Hydro-Québec by radio wave.

To Hydro-Québec's surprise, the change brought not accolades, but a public outcry. Citizens' groups were first to sound the alarm, claiming the radio waves emitted by the meters would create magnetic fields with "well-known" effects on human health." Some opponents of the meter went as far as claiming they would

Fig. 12-A: Hydro-Québec's new smart meters: much ado about nothing.

increase the risk of leukemia in children. The controversy snow-balled for two years, climaxing when Quebec's left-wing party Québec Solidaire demanded an outright moratorium on the new meters. The party based its demand on health concerns related to the meters, as well as the high fees Hydro-Québec was charging customers who refused to have the new meter installed in their homes.

Critics paid no heed to the fact that Italy had installed 27 million smart meters some 15 years ago, with no discernable impact on public health. The consumer protection organization *Protégez-Vous* finally deflated the hysteria in November 2012 when it published the results of an independent study done in collaboration with Université de Montreal's École polytechnique engineering school. The study found that the intensity of the radio waves from the smart meters was well under Canadian standards and that smart meters would contribute almost nothing to the radio frequencies already omnipresent in Quebeckers' lives as a result of Wi-Fi networks, computers and smartphones.

Hydro-Québec was perplexed by the intensity of the flak it got for the smart meters, and concerned it wouldn't have much leeway to make similar decisions in the future. When Éric Martel took up the position of President and CEO in July 2015, he swore that under his leadership Hydro-Québec would never again be a victim of false information campaigns. "The story completely spun out of control," he told Jean in an interview for *L'actualité* magazine in 2016. Martel had just spent months giving his employees and managers pep talks: all were mystified by the population's negative reaction to the smart meters. "We do some extraordinary things, but we really get beaten up in the media and on the web. I really think we do things better than we are given credit for."

Hydro-Québec has struggled with a persistent image problem for a number of years. Before Éric Martel's arrival, Hydro-Québec Distribution undertook a series of reforms to improve the management of its customer service, the quality of its communications during power outages, and its communication with customers in general. Among other things, it extended its customer service department's opening hours. Hydro-Québec now answers phone calls in the evening and on weekends—something unthinkable, even as recently as 2010.

In recent years, Hydro-Québec's Board of Directors realized that the company's image problem was also hurting business. Persistent criticism of Hydro-Québec created a climate of mistrust among customers and the general public, as well as something of a siege mentality among staff. At a certain point, this kind of situation can block a company's ability to make, or act on decisions. Or it can push a company to making self-defeating decisions. When Éric Martel became President and CEO, the Board of Directors' first order was to rebuild confidence among customers and the larger public. "We're not going to let anyone criticize us for things we do well, like the smart meters, or our rates," Martel said after taking up the job. He admitted, however, that Hydro-Québec bore some responsibility for the situation. "We were very shut off from

Fig. 12-B: A glass house. Every decision and every move Hydro-Québec makes is scrutinized by the media, the public and politicians.

the public. We have to be closer to people, we have to explain things better, answer questions."

Always Starting Over

So Éric Martel decided to dedicate his first months at the helm of Hydro-Québec to rallying his personnel. He invited his 500 top managers to monthly conferences and visited close to 5000 employees in regions throughout the province. The Martel gave up the private elevator the previous administration had taken over and made a point of eating regularly with employees at the cafeteria. He even pulled on the orange jumpsuit line workers wear, climbing up electric towers and onto service trucks.

From Day 1, Martel worked to put an end to Hydro-Québec's traditional aloof stance toward the public. He made sure employees answered the public's questions and responded promptly to complaints. "We won't be doing gymnastics anymore to try to fend off questions from the public or the media. We have nothing to hide. And if we can't explain something because of regulatory or contractual reasons, we'll just say we can't," says Martel.

Quebec's government-owned hydroelectric utility has made a big effort to modernize its communications department. It created a smartphone application customers can use to get up-to-date information on power outages. Customers can see a breakdown of their energy consumption in dollars and kilowatt-hours by logging onto the "My Customer Space" page on Hydro-Québec's website. Hydro-Québec bills also show the average outdoor temperature during billing period (so people can see why they used more or less electricity that month). Hydro-Québec also created a "Welcome to Hydro-Québec" campaign for television and the web, opened Twitter and Facebook accounts and hired a team of writers to moderate the sites.

In an effort to "loosen up" its traditional communications style, Hydro-Québec decided to publish questions and comments as they pop up, without correcting grammar or spelling mistakes. *"New house and it doesn't seem too well insulated. What do I do?"* asks Ginette from Thetford Mines. *"More advantageous to use my heat pump than the heater with temperatures under 15¢?"* asks Leo R. from Chicoutimi. Hydro-Québec answers all the questions—meaning *all*—often using a touch of humour, a first for the company. One Twitter message on September 12, 2018, read "Cold weather's on its way. Hydro-Québec better give us advice for saving electricity. Like masturbating somewhere besides the shower..." Jonathan Côté, Hydro-Québec spokesperson, answered this one: "If you want to cut your electricity bill, go ahead and do it in the shower, but use cold water." That would have been unthinkable five years ago.

Éric Martel's efforts have paid off. Between 2016 and 2017, the number of complaints about Hydro-Québec fell by 24 percent.[1] In October 2016, a report by the Saine Marketing firm noted, "an improvement in perceptions of hydroelectricity as a clean and renewable source of energy." The "feeling of [Hydro-Québec] being well managed" had also improved. At the same time, Saine Marketing found Hydro-Québec was generating what it qualified as "a high degree of negative noise" in the media.

When it comes to communications, Hydro-Québec is always starting over. No matter what the company does, or how its spokespersons explain things, Hydro-Québec is criticized for three things: its rates, power outages, and management. Hydro-Québec is not the only power supplier that struggles with its image. Power systems always have new customers to serve, households that move, new hookups to make, customers who lose their jobs, get into debt or are forced to live in poorly insulated apartments, others who forget to turn off their pool heater when they leave for holidays, municipalities that don't take care of vegetation that touches lines, politicians who make promises they can't keep, companies that need more megawatts or power, city dwellers who build chalets at the end of a road in the middle of the mountains, and plain whiners. In short, Hydro-Québec deals with life and all its complications, all the time.

But there are three main factors that contribute to Hydro-Québec's particular communications challenges.

First, because Quebec massively electrified homes and hot water heating, and because of Quebec's Nordic climate, Hydro-Québec is the single target for most of the antagonism related to energy costs in the province. Elsewhere in North America, like in Ontario, electric heating is rare, or at least rarer than in Quebec. So customers in other provinces pay two energy bills: one for electricity and the other for either natural gas or oil heating. In Quebec, three quarters of houses pay a single energy bill every month, to Hydro-Québec. If Quebec's government manages to increase the proportion of houses using electric heat from 73 to 100 percent, as it hopes to, the result will certainly even more customers complaining about their "hydro bill."

The second factor is the new phenomenon of the "reactive customer." Unlike in the 1960s, electricity customers today do not passively receive energy. Energy consumers today want to interact and participate in managing their energy consumption. This new "reactivity" isn't just about comfort, or even money, or personal interests of any sort. Customers are simply more conscious of how energy use impacts the environment, society and communities (see Chapters 13 to 15). This new reality opens Hydro-Québec up to even more criticism from the public.

Finally, there is the challenge created by a generalized lack of understanding of electricity among customers and the public. Unlike the cases of oil or natural gas, electricity can't be poured into a glass (or into a tank). It has no concrete existence (or none that's observable, anyway) and few people actually understand what it takes to control it or get it to customers. What's behind the meter is a mystery to most and this knowledge vacuum contributes to the general impression that Hydro-Québec's decisions are more or less arbitrary.

The Price of Electricity

Comparison of rates between 1963 and 2017

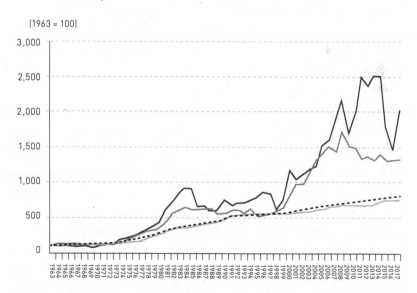

Fig. 12-C: Since 1963, Hydro-Québec's rates have risen at the same rate as the consumer price index. The same cannot be said about gas and oil prices.

Increases to electricity rates are always controversial. Faced with even a slight rate increase, people forget that Hydro-Québec has the lowest electricity rates on the continent. It's the Régie de l'énergie to make sure it stays this way. Since 1963, Quebec's hydroelectric rates have increased more slowly than inflation, on average, while the price of natural gas has increased 66 percent faster than inflation, and oil, three times as much over the same period (Fig.12-C). For four years now, Hydro-Québec has systematically kept rate increases below the rate of inflation.

Not only are Quebec's rates the lowest on the continent, its residential rates are actually lower than they should be, in theory. According to studies by Pierre-Olivier Pineau, Research Chair in Energy Sector Management at Montreal's HEC business school, the residential rate applied in Quebec amounts to about 85 percent of its actual cost price (Fig. 12-D). In other words, Quebec customers pay 15 percent less for electricity than what it costs to produce and supply it to them. Industrial customers generally pay 104 percent of the cost price and commercial, institutional and businesses pay around 130 percent. The pricing method called "cross-subsidization" (or cross-funding) is used by very few other electricity utilities on the continent.

In other words, the customer that Hydro-Québec "subsidizes" isn't usually the one people believe it to be. Cross-subsidization is an old practice put in place at the time of nationalization, and no Quebec government has ever been willing to question it since. If the Quebec government managed to electrify 73 percent of the heating in Quebec, and if half of the overall energy consumption comes from renewable energy sources, it's thanks to the cross-subsidization that principally benefits residential customers.

Since electricity is a force that's difficult to understand, Hydro-Québec will always face a serious challenge in the area of popular education. It only takes a few words in a conversation about electricity to see that even educated people fail to grasp the basics of the distinction between electricity rates, and the total price on their bill.

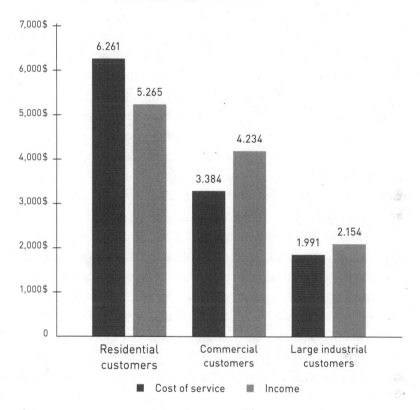

Illustration of cross-subsidization in 2017, in millions of dollars

Fig. 12-D: The subsidies aren't where they are assumed to be. Residential customers, on the whole, pay $996 million less per year than what it costs to supply them.

In customer discussion groups, Hydro-Québec constantly grapples with the problem. Customers participating say things like "I should never pay more for electricity since my thermostat is always at 20°C." Or "I don't care about the outside temperature. What I want to know is the temperature inside." Since 2018, Hydro-Québec's bill indicates the average outside temperature during the billing period on statements so customers can understand the effect of average temperature on actual cost, and why they are paying as much (or as little) as they do.

But that extra information doesn't change the fact that every rate increase is met with a chorus of protest, even when that increase is negligible (like the last one, 0.8 percent). Even if the increases are on average much smaller than rate increases for oil, natural gas or any other type of energy, the impact is felt disproportionately in Quebec because electric heating is so widespread.

"Don't forget that there are 500,000 low-income households in Quebec," explained Viviane de Tilly, Energy Analyst at the consumer rights' association Union des consommateurs. Viviane de Tilly knows her electricity: she worked in different jobs at Hydro-Québec for over 20 years before "going to the other side" in 2012. "There are people who have real trouble paying their electricity bill. When you are living on welfare, $1500 per year is a lot."

Although Quebeckers pay less for electricity than anywhere else on the continent, heating alone accounts for 54 percent of their average electricity bill, so it's not hard to imagine the impact rate increases have at the end of the month. "It's the bill we need to talk about, not the rate," says de Tilly. "New Yorkers pay much more per kilowatt-hour than Quebec, but they only consumer 3600 kWh per year. In Quebec, it's five times that amount. In British Columbia, only 10 percent of the population heats with electricity, the rest with natural gas."

Since electricity is less expensive in Quebec than elsewhere, and less expensive than natural gas or oil, Quebeckers' overall energy (electricity) bill should be lower than that of other geographic areas. But that's not necessarily the case. For starters, Quebeckers have been particularly prodigal when it comes to electricity consumption—they are considered among the highest energy consumers in the world. The reasons are numerous. Quebec has a high population of renters who often have to settle for housing that's poorly insulated and poorly heated (The Quebec government has allowed the use of wall heaters to proliferate; they are cheaper to install but much less efficient than central heating). That makes the bill at the end of the month that much higher.

However, this does have to be put in perspective. The average consumption for Quebec households is 17,000 kWh per year, which costs about $100 per month. Most households spend much more than that on Internet, phone and cable services. And since 2016, Hydro-Québec has significantly raised the level of consumption when the higher rates apply. The effect is that for customers with modest household incomes, a 0.8 percent increase is really a 0.1 percent increase, or 12¢ per month. "The 0.8 percent increase is the average for all categories of customers," says Marc-Antoine Pouliot, Manager, External Relations Strategy at Hydro-Québec Distribution. An increase of 0.8 percent translates into a rate increase of 23¢ per month for a small house of 111 m²; $1.60 per month for an average sized house of 158 m² and $2.44 per month for a large house of 207 m². Hydro-Québec has also implemented a number of programs to help low-income households when people find themselves unable to pay their bill; these households are encouraged to contact Hydro-Québec. Unfortunately, too many customers wait until the last minute to avail themselves of these services.

Hydro-Québec, of course, can't be held responsible for the way customers manage their finances, nor for customers' choices when it comes to heating, nor for poorly insulated housing or government housing policies. "It's not Hydro-Québec's job to redistribute wealth," concedes Viviane de Tilly. "We would like electricity to be less expensive. But it's not up to Hydro-Québec to solve the problems of low-income families. There has been some progress, but there could be more."

Who's Hydro-Québec Working For?

Generalized confusion about Hydro-Québec's vocation also compounds the perception problem about rates. Does Hydro-Québec work for its customers? Or for taxpayers? Or for the government? The answer is actually all three.

Yet misunderstandings pop up constantly about the nature of Hydro-Québec's vocation. Among the most common complaints: why don't Hydro-Québec's profits translate into lower rates? When Hydro-Québec's CEO has announced a promising new export contract with Massachusetts, or Ontario or somewhere else, he has had to answer the same question: "If Hydro-Québec is making more money, why aren't Quebeckers paying less for their electricity?" Yet the question is not even up to Hydro-Québec to answer. It falls on Hydro-Québec's single shareholder, the Quebec government.

Since the Hydro-Québec Act was updated in 1981, Hydro-Québec's mandate has not been *exclusively* to supply electricity "at the lowest rates consistent with sound financial management," (Article 22 of the original law, passed in 1944). In 1981, the Quebec government transformed Hydro-Québec into a shareholding corporation, and gave it a commercial mandate. Since then, Hydro-Québec's single shareholder, the Quebec government, expects to be paid a dividend at the end of the year. So for the last 40 years, Hydro-Québec has served three masters: its shareholder, taxpayers and customers. What's surprising is that after 40 years, journalists still haven't understood this situation.

Hydro-Québec also went through another change at the end of the last century. Since 1996, neither the government nor Hydro-Québec has decided what electricity rates will be. That's the job of Quebec's Régie de l'énergie. Each year Hydro-Québec submits a report to the Régie with details about its assets, debts, expenses and sales for the year and its projected dividend for the coming year. Before consenting to a rate increase (or not), the Régie analyzes the figures for eight months. Then on March 1, every year, it announces what increase will be allowed (if any).

The Little Outage That Grew

After rate increases, power outages are the next topic that sparks emotions about Hydro-Québec. While rate increases are never welcome news, outages, especially large-scale ones, generate a more complex reaction from the public, one that mixes fatalism and human solidarity. A case in point: Quebec's 1998 ice storm. "In the 1970s and 1980s, there were some general power outages that provoked a lot of criticism and changed attitudes toward Hydro-Québec. This was the case particularly for the big outages of 1988 and 1989," explains André Bolduc, a former economist at Hydro-Québec who has written many books on the company. "But the ice storm of 1998 had the opposite effect, at least initially." In January of that year, 100 mm of ice built up over five days of freezing rain, destroying hundreds of electricity towers and thousands of poles. Up to 1.3 million customers (3 million Quebeckers in all) were plunged into darkness, after which the Canadian Army deployed 15,000 soldiers in the biggest rescue operation in Canadian history. Hydro-Québec's employees worked feverishly to get the system up and running. In one case, fearless line workers jumped from helicopters onto electrical towers straddling the river between Kahnawake and LaSalle to keep the network feeding Montreal from failing. "You could say there was a sea change in popular opinion after the ice storm. It put a human face on Hydro-Québec," says Bolduc.

Hydro-Québec's "planned service interruptions"—when Hydro-Québec has to temporarily cut the electric current to repair a line or a transformer—are another matter altogether. Customers can be stoical up to a point, but lose patience when power cuts last longer than predicted.

"There is, literally, always an outage somewhere in the system, whether it's caused by a falling branch, a car that hit an electric pole or a planned interruption to service that goes on longer than expected," explains Patrice Richard, Manager, System Operation Activities at Hydro-Québec.

One of the significant changes that have taken place at Hydro-Québec, but has not been reported in the media, was the creation of the Centre de gestion des activités de distribution (Centre for managing network operations, CGAD, described in Chapter 1).

Fig. 12-E: A particularly destructive microburst hit the Montreal neighbourhood of Notre-Dame-de-Grâce in the summer of 2017. Emergency teams have to be on site at a moment's notice.

Its mandate is to coordinate Hydro-Québec Distribution services and territories, and monitor the quality of services on a continual basis. Patrice Richard invited us to visit the CGAD's crisis centre, a glassed-in room with walls covered in screens. Beside a TV streaming Radio Canada's news channel, an electronic table shows all the power outages on the system, updating their progress every minute. On the afternoon we visited the CGAD, there were 30 outages affecting some 1200 customers in Quebec. Richard said it was a pretty normal day except for one major outage affecting 600 customers. The table also showed planned interruptions and the number of customer service calls to report issues like, "A branch fell on a line," or "A lightning bolt hit a pine tree in front of our house."

The objective of the CGAD, created in 2013, is to improve management of the system and customer service. Hydro-Québec Distribution has five administrative territories. Normally, each one takes care of its own business. But the CGAD intervenes when a region is overwhelmed and can't respond to an emergency on its own—a situation that occurs an average of 30 times per year. "In 2010, when a storm front tore things apart, it took us 36 hours to restore service. Now we manage to do it in 25 hours," says Richard. One way the CGAD has managed to pull this off is by improving prevention methods. "When we see a storm front coming, we know there will be outages. So we send teams to specific areas before it hits. We call back employees on holidays, or we ask employees about to head home for the day to stay at work longer."

Éric Martel regularly reminds Hydro-Québec staff and management that adopting a "customer approach" means they have to stop thinking like a monopoly. The idea was already sinking in before the arrival of the new CEO, notably at Hydro-Québec Distribution. "With the advent of energy self-generation, the possibility of our customers going elsewhere is very real. In five or ten years, our customers might actually have the choice of whether or not to buy electricity from us, so we're going to have to do a better job if we want to hold on to them," predicts Richard. His message to employees is: you can't be satisfied with just "getting the job done." You have to think in function of customers' needs."

Patrice Richard is not the slightest bit wary of talking about Hydro-Québec Distribution's weak points. "We are very good at handling emergencies and power failures, but we have a lot of work to do when it comes to planned service interruptions," he admits. He cites the case of Quebec City's Saint-Roch neighbourhood where an operation to repair underground equipment in 2018 meant Hydro-Québec had to cut power to 30 residences from 8 a.m. to 4 p.m. It was -25 °C outside that day. Customers had been forewarned via a telephone campaign and the work started at the scheduled time. But at noon, the team informed their overseer that the job was taking longer than expected. They thought they would be working until 8 p.m. The problem? The overseer didn't transmit the information to management. "We heard about the delay from a tenant who called our customer service department to complain. He had not learned about the delay from us, but from one of the workers on the site. Luckily it didn't make the evening news. But we don't want the press or Hydro-Québec's upper management to learn about problems before we do. When something happens, the CGAD needs to know about it so it can organize a response," says Richard. "It sounds self-evident, but it's actually not easy to get everyone thinking this way."

The Right Hand Following the Left

Quebeckers have the stubborn, ingrained idea that Hydro-Québec is a huge, poorly managed monopoly. "People are proud of what Hydro-Québec did in the past, but in customer surveys they only give the company 6 out of 10 for management, even though they're not sure why," says Serge Abergel, Manager, Public Affairs and Media at Hydro-Québec.

"Until 1970, Hydro-Québec's reputation was squeaky clean. It couldn't do any wrong. No one questioned how it was managed. But that idea changed over the course of the 1970s when the company's choices started becoming controversial. Quebeckers were better educated by then, better informed, and they were becoming more conscious of the different impacts Hydro-Québec's projects could have," explains André Bolduc. The James Bay Project, which began construction in 1971, was a major factor in this change of attitude. "Ecological considerations had become more important by that time, as well as concerns over relations with Indigenous communities in Quebec. No one had given them much thought before that. That's when journalists and commentators started to refer to Hydro-Québec as a state within a state."

Although the 1998 ice storm initially had a positive impact on Hydro-Québec's image, the wave of solidarity didn't last. Popular thinking soon returned to what it had been before the storm: a sort of general distrust of Hydro-Québec's motives, reinforced by a vague but tenacious idea that the company was a poorly managed behemoth.

Though things could always be better, it's somewhat naive to think a company of 20,000 employees will operate like a perfectly calibrated machine at all times. Like in any large corporation, decision makers in one division don't always know what their counterparts in other divisions are doing, or how they can help them. Turbulent dealings between different services can at times rise to a surreal level. And while customers may be frustrated by poor communication from Hydro-Québec Distribution, they should know that Hydro-Québec Distribution sometimes has the same problems with its own supplier, Hydro-Québec TransÉnergie. "You'd think common sense would ensure cooperation, but the reality is, it doesn't always happen," admits Patrice Richard.

Managing Hydro-Québec is also complicated by the fact that some of its "decisions" actually come from its single shareholder, the Quebec government. The government makes demands and determines the orientations of the government-owned company, in some cases even dictating specific actions. Sometimes things turn out well. Sometimes they don't. Between 2004 and 2007 the government forced Hydro-Québec to call for tenders from the private sector for 41 contracts, for a total of 3900 megawatts.[2] Hydro-Québec's preference would have been to develop at least part of the capacity internally, as it did for dam building. Calling for private bids meant Hydro-Québec had to give up control over managing the windmills, so it would essentially be forced to purchase its wind energy generation all the time. The government's hope was to use wind power contracts to spur regional economic development in the Gaspésie and the Bas-Saint-Laurent region by getting local companies involved.

But things didn't work as expected. Demand for electricity started to stagnate in Quebec in 2007, as it did in all developed countries, but the wind power contracts with private developers had already been signed. Hydro-Québec would end up paying for alternate energies that only added to its growing surplus. And *voilà*: Hydro-Québec was locked into deals forced upon it by its only shareholder.

In normal times, when the government wants to stimulate regional development, it uses its own budget, and taxpayers foot the bill. However, in the case of the windmills, the government handed the challenge to Hydro-Québec, forcing it, in the process, to come up with the financing on its own by raising rates for customers (which it can't unilaterally do). And not only that. Hydro-Québec then had to solve the technical challenge of injecting 1900 intermittent wind turbines into its system without disrupting power supply. For Quebeckers, the real cost of this adventure will no doubt show up in rate increases in the near future.

Chapter Thirteen

Visible from Space

Like almost all the astronauts who get to the International Space Station, Canada's David Saint-Jacques probably took a picture of the "eye of Quebec" when he made his first walk in space in April 2019. The spectacular reservoir behind the Manic-5 dam, a 100-kilometre-wide ring with an island in the middle, is the fifth-biggest meteorite crater on earth. At 330 km in altitude, astronauts can't see the enormous Daniel-Johnson Dam that holds the water, but they can see the reservoir.

The Manic-5 reservoir perfectly demonstrates the paradox of Quebeckers' energy choices. Of all forms of energy, hydroelectricity is among the least harmful to the environment and contributes the least to climate change. However, in the form of mega-dams like Quebec's, hydroelectricity leaves the most visible trace on geography and nature of any energy source—and by far.

Quebeckers are always going to have to live with the physical consequences their energy choices entailed. To build large dams, big reservoirs and long transmission lines, civil engineers construct new environments by first destroying old ones. It's not the same scale of destruction as an atomic bomb reaped on Hiroshima, or the kind of disdain for nature that left a desert of oil sands around Fort McMurray. Still, to build dams, Quebec flooded thousands of hectares of land and bogs, transformed rivers

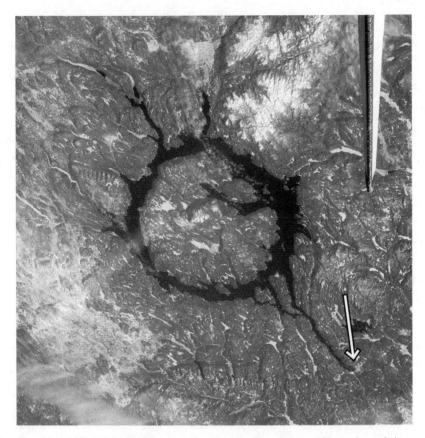

Fig. 13-A: The Manicouagan crater viewed from space. Gigantic as it is, the Daniel-Johnson Dam (arrow) is not visible from space. The wing of the Columbia Space Shuttle is visible in the upper right corner of the photo.

into lakes, razed whole forests, disturbed habitats and destabilized entire communities. Add new power lines to the mix and the overall impact of dams on Quebec's landscape becomes enormous.

By the very nature of its work, then, Hydro-Québec must constantly manage environments—along with public perception. Because in the 75 years since Hydro-Québec was founded, the environment has become the focus of both complicated scientific work and highly charged emotions.

It will probably come as a surprise to many to learn that Hydro-Québec is, on the whole, widely respected for its environmental practices. The magazine Corporate Knights, which describes itself as dedicated to "clean capitalism," rated Hydro-Québec first among the "Best 50 Corporate Citizens in Canada" in 2018 for getting "99.5 per cent of its production from renewable hydroelectricity."

Environmental concerns are so widely embraced today that it's easy to forget how recently they entered the public's consciousness. No one talked about the environment in the 1940s when Hydro-Québec was founded, or even in the following decade. Even two generations ago, ecological concerns were still limited to the counterculture movement and a handful of avant-garde scientists. Efforts to reduce pollution only started at the end of the 1960s. "When I arrived at Hydro-Québec, at the beginning of the 1970s, the word "environment," in today's meaning, was barely used in English and not at all in French," recalls Roger Lanoue, former vice-president of research and strategic planning from 1999 to 2004 and co-chair of the Quebec Commission on Energy Issues in 2013. It wasn't until the 1970s that the concept of greenhouse gases, which originates in the 19th century, broadened the environmental discourse to include climate change. The notion of "sustainable development"—meaning economic development projects also had to meet ecological imperatives and gain social acceptability—soon followed.

Hydro-Québec has adapted to each of these developments, in some cases even anticipated them. However, its environmental work has been carried out in a context of almost permanent controversy. Since hydroelectricity harms the environment (even if only visually), environmentalism has had an impact on Hydro-Québec's work as well: the company has had to cancel numerous projects because of environmental concerns, whether legitimate or not.

Learning As They Go

Hydro-Québec today has 180 biologists, geographers, lawyers, archaeologists, oceanographers and chemists working on hundreds of environmental impact studies. They create indexes of fish, algae and water birds, study peat bogs marshes and compile tens of thousands of pages of reports. Hydro-Québec has to respect dozens of federal and provincial laws in its environmental practices. Just for phase three of the James Bay Project (on the Eastmain, Sarcelle and Rupert rivers), Hydro-Québec had to obtain close to a thousand authorizations from the federal and Quebec governments and the Cree community.

Things haven't always been like this. A year before the Quebec government announced its intention to develop the hydroelectric potential of James Bay in 1971, Hydro-Québec had created what it called a "committee for the protection of the environment." The committee's responsibility was to make sure Hydro-Québec was doing everything it could to reduce pollution, which was the main environmental concern at the time. Hydro-Québec created an official Environment Department in 1973.

Roger Lanoue, who started working for Hydro-Québec after he finished a master's in public administration at Cornell University, witnessed the birth of environmental consciousness at Hydro-Québec at the beginning of the 1970s. Protests against the James Bay Project had started. The Cree peoples were doing everything they could to put a stop to the project, including citing its potential environmental damage. Hydro-Québec had to react. "Starting in 1968, the United States had environmental protection legislation. In Canada, there was almost nothing at the federal level and in Quebec, there was nothing at all at the provincial level," recalled Lanoue. Quebec passed a law related to the quality of the environment in 1972, but it concerned strictly air and water pollution. "At the time Hydro-Québec was still the only organization in Canada with any environmental regulations of any sort."

Hydro-Québec became one of the first companies in North America to produce environmental impact studies. "When we started doing them, it was like doing a dry run to practise, before formalizing the procedure. We had the feeling these reports would soon become very normal procedure," says Lanoue.

Roger Lanoue and his colleagues were working in a technical and juridical vacuum, a situation that lasted for years, since Hydro-Québec's Environment Department was created six years before Quebec's Environment Ministry. Hydro-Québec's Environment Department also produced its first impact studies long before Quebec's Bureau d'audiences publiques sur l'environnement (Office of public environmental hearings, BAPE) was established in 1978. The BAPE didn't have its first hearing until 1980. Hydro-Québec also created its own Code de l'environnement in 1981, three years before the provincial government had an environmental policy. In 1985, Hydro-Québec had a program to highlight environmental issues that stipulated 1 percent of the price tag of projects should return to the community. That was two years before the United Nations Brundtland Report introduced the idea that sustainable development should be accompanied by measures to support communities. In 1995, Hydro-Québec formulated its first report on sustainable development, 11 years before the provincial government created a law on sustainable development.

The James Bay hydroelectric project unfolded just as environmental concerns were becoming more mainstream in Quebec and elsewhere. Although Hydro-Québec produced environmental inventories during the first phase of the project, no comprehensive environmental impact studies were performed as they are systematically done today. "The first "environmental report" on the project, in 1971, was a review of the literature on the ecological impact of the project made without visiting the site," says Anne-Marie Prud'homme, Scientific Communicator at Hydro-Québec. Hydro-Québec's first real environmental impact study was carried

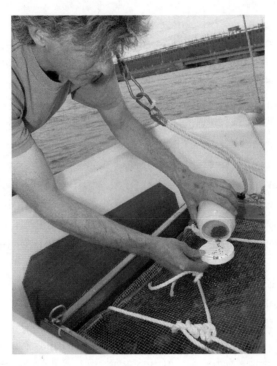

Fig. 13-B: A biologist handling sturgeon eggs. Since the 1970s, Hydro-Québec has carried out detailed environmental impact studies for all its projects—future and past.

out for the 528-megawatt Outardes-2 generating station commissioned in 1978.

Things changed radically after that. Ever since phase two of the James Bay project, Hydro-Québec does a thoroughly documented environmental impact study for all its projects. Public hearings are also held for each project. Hydro-Québec does not go as far as naming individual fish, but close to that. In 2004, the environmental impact study Hydro-Québec produced for the Eastmain and Rupert rivers was presented in 15 volumes.

Hydro-Québec's Environment Department has also done reviews of all finished projects to make any corrections necessary. For example, in 1929, Hydro-Québec's predecessor, Montreal Light

Heat & Power, built a generating station on Rivière-des-Prairies, north of Montreal, without realizing the area was one of the most important spawning grounds for lake sturgeon in southern Quebec. Today, there are only a few Hydro-Québec dams—including the oldest ones—that don't have fish ladders or other devices to allow the movement of fish species. This, of course, requires producing thorough studies on the behaviour of different species of fish. Some fish, like American shad, prefer travelling down spillways. For the shad, Hydro-Québec uses ultrasounds to deflect the fish away from turbines and guide them to the spillways, which they clear without problems.

The ecological science of natural habitats is at least as complex as the physics of electricity—maybe even more so, because of the huge range of habitats and the need to include a variety of fields to study them. Depending on the species, these can range from physics to chemistry and biology and even include human sciences like sociology and economics. Hydroelectricity is at the junction of many—if not all—fields of science, and growing environmental consciousness adds to the complexity. The challenge has become more complicated since environmental protection, limited to ecologists as recently as the 1970s was, is now a universal value shared by almost all Hydro-Québec customers, as well as those in export markets.

So Hydro-Québec today has to arbitrate conflicting opinions, including some held by groups who, legitimately or not, have used environmental arguments to add weight to other claims. For example, the Cree communities in Quebec succeeded in getting a major electricity export contract to the State of New York cancelled in 1992 thanks to help from the U.S. ecological lobby. The cancellation of the contract spelled the end of the Great Whale project. (U.S. ecologists simply brushed off the fact that the only viable alternatives for energy at the time were a conventional hydrocarbon thermal energy generator or nuclear power.) Citizens groups and Indigenous communities forced Hydro-Québec to abandon a project on the Ashuapmashuan River, in Lac-Saint-Jean, the same year. The environmental card does have its limits. In 2018, the Innu population in Pessamit (formerly Betsiamites), 50 km southwest of Baie-Comeau, argued to the Boston press that a proposed electricity export contract to Massachusetts would be a direct threat to their way of life. The argument didn't succeed in stopping the project.[1]

On the whole though, the criticism Hydro-Québec has been fielding from the U.S. for the last 30 years has been more culturally motivated than scientifically based. It began as a question of defending the rights of First Nations peoples, then splintered across a range of environmental concerns. According to Roger Lanoue, the main motivation behind U.S. criticism of Hydro-Québec has always boiled down to good old-fashioned protectionism. "Americans have always found reasons to argue that the electricity coming from anywhere else is not environmental. The real reason they oppose it is that Quebec's hydroelectricity competes with their production, so as far as they are concerned, any argument against it is fair."

Of course a number of associations in Quebec have also called Hydro-Québec to account, including the Fondation Rivières (Rivers foundation). Founded in 2002 by Quebec actor Roy Dupuis, filmmaker Michel Dupuis and engineer Alain Saladzius, the non-profit association's first campaign was to oppose a Quebec government program to encourage private energy generators to build and run mini hydroelectric generating stations of less than 50 megawatts. The association was also among the most tenacious voices protesting Hydro-Québec's decision to build the La Romaine hydroelectric complex. It continues to argue against the complex to this day, even though it is almost completed. "We agree that hydroelectricity is less destructive to the environment [than other energy sources], but it still has an environmental impact on erosion, on wildlife and on the rate of water flow. Artificial water management destabilizes the natural balance of a river, the presence and quality of nutriments in the water, everything," says Alain Saladzius, Co-founder and President of Fondation Rivières and a water management consultant who worked in various government ministries in Quebec for 32 years.

Although for decades Hydro-Québec did consider environmental concerns mostly as obstacles it had to surmount, the company now considers its scrupulous approach to environmental management to be one of its best calling cards, especially in the context of climate change. The fact that states in New England, and Ontario are committing to reducing their dependence on fossil fuels and nuclear energy, combined with the push toward alternative energies (see Chapter 9), may also make it easier for Hydro-Québec to justify building new dams and transmission lines.

Since 2009, states in the North East—including Connecticut, Delaware, Maine, Maryland, Massachusetts, New Hampshire, New Jersey, New York, Rhode Island and Vermont—have all committed to targeting the energy sector to reduce their GHG emissions. In 2016 and 2017, Vermont, Massachusetts and the State of New York adopted specific measures to force their energy suppliers to include renewable energies in their supply mix. Those rules were what led to Hydro-Québec's 2018 export contract with Massachusetts. There is good reason to believe there will be more of these in the future.

To manage perceptions, Hydro-Québec has learned to dissociate export contracts from local construction projects. In the past, the company used to announce, "We'll build a dam right away and sell the electricity to Americans." That approach gave opponents to the projects and competitors both time to organize a response. In the controversy over the Great Whale project, the Cree went as far as inviting Robert Kennedy Jr. to paddle down the river in a kayak, leading to a public relations disaster for Hydro-Québec. The change in approach shows in Hydro-Québec's recent export contact with Massachusetts: in the past, the company would have used the contract to justify building a dam complex, like La Romaine. Now the two questions are completely separate.

It will be interesting to see what happens with export projects under Quebec's new premier, François Legault. Legault has shown signs he wants to use electricity exports as a pretext for launching new dam projects in Quebec. He has a new argument: "There are still groups like Fondation Rivières who oppose building new dams. But what must be understood, and what we must explain, is that there is no wall between Quebec and U.S. states. Compared to the effects of gas and coal generating stations, harnessing the power of a river has much less impact on the environment. If the states in the North East replace their coal and gas stations with hydroelectricity, the whole planet will win, and so will Quebeckers," says Legault. "It could be Quebec's biggest contribution to reducing GHG emissions on the planet." Future debates over this will be fascinating.

Clean, But How Clean?

Debates over the environmental impact of Hydro-Québec's projects would spark less emotion if one simple truth were recognized: "clean" energy, or energy that has no effect on the environment, does not exist. All forms of energy have to be produced, transported and stored before being used and these activities all have environmental impacts.

But the combined effects of different types of energy production can be measured and compared. The most universally used standard is GHG emissions or pollutants. But there are other ways, like the effect of different types of energy on plant and wildlife or on neighbouring communities. It's even possible to tally up qualitative aspects like the effect on landscapes, for example (though in these aspects, the line between science and politics becomes fuzzy).

It's hard to find a more specific analysis for comparing the GHGs and carbon footprint of different energy sources than

GHG Emissions - Power Generation Options

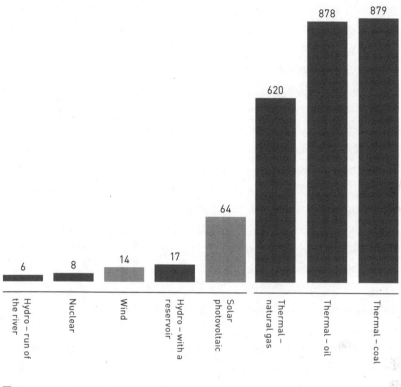

Uninterrupted power suppy
Intermittent power supply

Fig. 13-C: Some kinds of energy are cleaner than others.
Hydroelectricity produced from water flow (not a reservoir) only
emits six grams of CO2 equivalent per kilowatt-hour. That quantity
is four times higher, on average, for hydroelectricity produced using
reservoirs. But this is still four times less than solar energy and
50 times less than gas and coal.

the one done by the International Reference for the Life Cycle
of Products (CIRAIG), a Quebec research institute that includes
scholars from all Quebec universities. By analyzing the life cycles
of different means of energy production, CIRAIG calculates the

climatic effect each type has in equivalent grams of CO_2 per kilowatt-hour of energy produced (see Fig. 13-C). Nuclear energy produces 8 grams of CO_2 per kilowatt-hour, while coal produces some 879 grams. In other words, if ten 100-watt light bulbs ran for an hour on coal, the effect would be almost a kilogram of CO_2 released. Oil does a little better, producing exactly 1 gram less of CO_2. Between the two extremes, natural gas produces 620 grams of CO_2. Solar produces 64 grams, while wind power is much lower, at 14 equivalent grams of CO_2 per kilowatt-hour.

CIRAIG can even draw a nuanced picture of CO_2 production according to specific conditions. The performance of hydroelectricity, for instance, is variable. A dam with a reservoir full of water (a small reservoir) produces barely six grams of CO_2 per kilowatt-hour. However, when the generating station is coupled with a big reservoir, the amount increases to 17 grams, or slightly more than wind power, but markedly less than solar energy. If the 23 isolated communities in Quebec that are now dependent on diesel or oil generators are included, the global impact of Hydro-Québec is 19 grams of CO_2 per kWh.

"Hydroelectricity, as it is produced here, is one of the cleanest sources of energy on the planet," says Normand Mousseau, a physics professor at the Université de Montréal who was co-chair of the 2013 Commission on Energy Issues for Quebec. Mousseau is surprised that solar and wind power are considered "green energy sources" in popular thinking, while hydroelectricity isn't. "Solar is relatively green but much less so than hydroelectricity. Analyses have to take into consideration the materials used for solar. In the case of hydroelectricity, it's cement. Making solar panels requires the extraction of silicon and other rare metals."

"Yes, territories have been sacrificed to build dams. But that would be the same for solar or wind power," Roger Lanoue tells us. For example, to replace the La Romaine Complex with wind power, Hydro-Québec would have to build wind mills covering 500 square kiometres of land. That would mean building roads and power lines as well. And to top it all off, Hydro-Québec would still need to have a dam, or a natural gas generating station to balance the wind production, which is intermittent at best.

"No alternative to hydroelectricity is as clean," concludes Normand Mousseau. "Which doesn't mean it's perfect."

Hydro-Québec has been carefully documenting its own GHG emissions for 20 years now, sending numerous teams across Quebec's territory and offering technical and financial support to university researchers, including letting researchers use their helicopters. Just two generations ago, Hydro-Québec didn't even bother cutting down trees before impounding the reservoirs. But it has been a long time since Hydro-Québec flooded virgin territory without doing an in-depth environmental impact study first. Today, where possible, trees are cut and removed from the zone. However a lot of vegetation remains both above and below water, not to mention peat bogs. When it decomposes that vegetation releases enormous amounts of CO_2 under water, particularly in the first year (the level of CO_2 returns to normal quite quickly).

The quality of impact studies has also improved a lot. Hydro-Québec can now study the GHG emissions of a reservoir to compare the quantity of GHGs released by forests and peat bogs before and after construction. Researchers have different methods for studying CO_2 and methane (CH_4), including taking samples using "floating chambers" mounted on rafts.

A 9000-year-old peat bog can store thousands of tonnes of CO_2 but the second it is flooded these gases start to be released bit by bit. Studies on Quebec's Eastmain River have shown the effects of this are significant, but time-limited. "When the bog is first flooded, GHG emissions are relatively high, but they return to normal after 5 or 10 years," said Michelle Garneau, Director of the Centre de recherche en géochimie et géodynamique (Centre for research in geochemistry and geodynamics, GEOTOP) at the Université du Québec à Montreal, who has carried out many research projects in peaty habitats.[2]

Though GHGs have become a universal concern in environmental protection, they are far from Hydro-Québec's only concern. Hydro-Québec deals with a wide array of environmental issues, from how to properly clean up old sites to evaluating the impact of new dam construction on mercury levels, or more generally on wildlife and humans, or biodiversity in general.

While not exclusive to the hydroelectricity industry, Hydro-Québec has its work cut out for it cleaning up old work sites from the 1970s. In the autumn of 2018, Ministry of Transport workers were drilling near Manic-5 when they discovered a former cement factory. When work on the generating station was finished, Hydro-Québec workers at the time simply buried the factory. "The tools were still on the table, right where they left them. That's the way people did things back then," says Francis Labbé, Hydro-Québec spokesperson. "Today, of course, we analyze sites to check for contaminants."

The cement factory at Manic-5 is not an isolated case. In 1998, Hydro-Québec identified some 1800 cases of buried waste on its site, most of which had been discovered by Cree trappers. The sites were mainly from Hydro-Québec's "glory days" building mega projects, from the 1950s to the 1980s. The contractors Hydro-Québec hired back then were no more concerned about environmental issues than the rest of society—in other words, not at all. They were just in the habit of burying old fuel barrels, even when they were not empty. Among the 1998 discoveries, workers uncovered some 2000 barrels buried close to a former cement factory that had been buried as is, just like the one at Manic-5. Since 1998 Hydro-Québec has spent $12 million cleaning up old sites, but there are still 800 left.[3]

Since the 1970s, Hydro-Québec has been closely monitoring mercury levels in its reservoirs by comparing them to natural habitats—mercury is present in all aquatic habitats and concentration levels tend to vary by species. In the last 40 years, Hydro-Québec has tested tens of thousands of samples of human hair, but also samples from wildlife, especially fish. "To our knowledge, there has never been a case of mercury poisoning caused by eating fish," says Anne-Marie Prud'homme. "The highest concentrations of mercury were measured in a fisher who was a big consumer of fish that came from a natural lake, not from a reservoir. The concentration was at a level considered safe." In other words, we're not talking about anything as severe as cases of Minamata disease, the neurological syndrome first discovered in post-war Japan that was caused by the release of methylmercury in industrial wastewater.

In nature, mercury levels can vary from one species to another. Lake whitefish and speckled trout absorb little mercury, while pickerel, pike and lake trout absorb higher levels. Concentrations can reach a level where it's sometimes necessary to limit consumption of certain species, even if they are caught in natural lakes.

"A lot depends on whether fish are predators or not, insect-eating or not, groundfish or not," explains Prud'homme. "We study the mercury problem from all angles, including its effect on the size of a species, its growth or population size. We even studied the diet of human populations living in the areas surrounding our projects. At La Romaine, for instance, we know that 1 percent of the diet of people living there comes from the territory. We haven't detected any effects on human health anywhere in Quebec in the last 50 years, even near our reservoirs."

As for wildlife, the picture is more nuanced and depends on the species in question, as well as the geography of each hydroelectric station. Building a dam and filling a reservoir necessarily brings a reorganization of the surrounding habitat. This process favours the proliferation of certain species but contributes to scarcity among others. These days, hydro dam managers take into consideration what species live both up and downstream and in the surrounding area. Dams affect some species more than others, but the overall environmental record—and even the records of those records, adding up to tens of thousands of pages of studies per year—have shown that the overall effect of hydroelectric dams on biodiversity is zero.

Furthermore, the conditions for operating hydroelectric generating stations are narrowly defined. The object of the dam is, of course, to generate electric power, but dams cannot be operated to the detriment of the facility itself, of the surrounding population, of any of the navigation activities or of the preservation of the environment. It's common sense. Treaties, laws and contracts regulate the conditions for operating dams. Biologists, for instance, have their say in how fish species are managed. Certain species tolerate strong water currents during some seasons but not others, especially during spawning season. Depending on the time of year, generation can be slowed to allow fish or other animals to pass through or to facilitate their movement. A large part of the flow of the Rupert River goes north to the La Grande, so Hydro-Québec and local Cree communities agreed on a procedure to simulate natural variations in the current. The rate of flow of the Rupert River is 127 cubic metres per second. To simulate spring floods, Hydro-Québec releases 416 m³ of water per second over 45 days in the spring and 267, again, in October over a period of 20 to 30 days. Hydro-Québec also invested several hundreds of millions of dollars to build dykes and erect other obstacles in an attempt to recreate the conditions of river rapids even when the rate of flow is decreased.

The list of measures being taken across Hydro-Québec's territory is long. It includes projects to protect sturgeon in areas including James Bay. At La Romaine, where the situation of Atlantic salmon was already precarious before the dam project, the company created new habitats on nearby rivers. "We're following our different projects closely," says Anne-Marie Prud'homme. "We know there will be neither loss of biodiversity nor the disappearance of any plants or animals."

The effects of power lines are also more nuanced than generally believed. Like forest fires, which are known to favour the proliferation of certain animal species, particularly large mammals, power lines can produce the same effect, while, of course, emitting fewer GHG emissions and causing less pollution than a fire.

That said, the principal concern with power lines today is their effect on the landscape, which brings us to the question of how Hydro-Québec is handling the explosive question of social acceptability for its projects.

Chapter Fourteen

Not in My Back Yard

"The new line is going to destroy the economy of our village!" In the spring of 2016, Julie was interviewing Lisette Lapointe, spouse of late Quebec premier Jacques Parizeau. Lapointe was mayor of the picturesque municipality of Saint-Adolphe d'Howard, a haven for outdoor-lovers in Quebec's Laurentians, and the spokesperson for a drawn-out campaign by villagers to prevent Hydro-Québec from building a 120-kV power line through their municipality. The ferocious battle of Saint-Adolphe-d'Howard had been making headlines for several years by that time. The village's residents protested at every public hearing and even produced a professional video to document their struggle. "All we have here is our village and the mountains," Ms. Lapointe told Julie. "Tourism is our only industry. The line is going to ruin us."[1]

The interview was Julie's first real brush with the issue of "social acceptability." One of the three pillars of sustainable development, it's not a new concept. Ever since the Brundtland report was submitted to the United Nations in 1987, development projects have been required to satisfy economic and environmental criteria, but they must also be accepted by the community where they will take place. However, while the principle was established decades ago, it has only become a really hot issue in the last decade, during which time Hydro-Québec has had to learn to manage it.

Fig. 14-A: In Saint-Adolphe-d'Howard, a high-voltage transmission line that would pass through 13 km of the municipality provoked a wave of protest from the day it was announced. Neighbouring municipalities were more open to accepting 28 km of the line.

Hydro-Québec has been managing social acceptability for its projects for many years. To some extent, it learned the ropes of social acceptability while managing environmental concerns (Chapter 13) and relations with Indigenous communities in Quebec (Chapter 15). As for social acceptability, Hydro-Québec has an elaborate consultation process in place that it has used repeatedly over the years.

The stakes for social acceptability are high for Hydro-Québec. One "no" from a single community can jeopardize an entire project. Social acceptability, for that matter, has been a big problem for Hydro-Québec's export contract to Massachusetts (worth roughly $10 billion over 20 years). The State of New Hampshire rejected the original path proposed for the transmission line, through the White Mountain National Park. (Hydro-Québec's American partner Eversource was responsible for the U.S. section of the line.) Fortunately, Hydro-Québec had prepared two alternative routes, one through Maine and one through Vermont. But social acceptability could be an obstacle for those routes, as well.

In fact, social acceptability comes up in just about every Hydro-Québec project. In a normal year, Hydro-Québec builds between 200 and 400 km of new lines, and they are rarely welcomed with open arms by the local populations. However cases of total deadlock are rare. For every situation where there's no possibility for compromise or negotiation, there are 100 other cases where Hydro-Québec manages to reach an agreement by making adjustments.

Hydro-Québec was completely caught off guard by the scale of the media attention and the intensity of debate over the line running through Saint-Adolphe-d'Howard. In some way, Saint-Adolphe-d'Howard was first. But it won't be a last. Social acceptability has become a much more important issue over the last 10 years, and is certain to gain importance in years to come. "We're living in an era where companies need social support to pursue their activities. There's no way around it. It will become more and more imperative with time," says Sophie Hamel-Dufour, a sociologist who specializes in public consultations and a former analyst at the Bureau d'audiences publiques sur l'environnement (Office of public hearings on the environment, BAPE). "There's a shift underway. The public looks at public administration with a much more critical eye than in the past. People are openly expressing their loss of confidence in politicians and the state."

Saying attitudes toward hydroelectric infrastructure in Quebec have changed is an almost laughable understatement. "Back in the 1960s, engineers in Quebec described the electric towers holding 735-kV transmission lines as little Eiffel Towers. They thought they were beautiful! From the point of view of mechanical engineering towers are fantastic creations, but no one would call them beautiful today," concedes Roger Lanoue. "Today Hydro-Québec has to consider not only the biophysical impact of its projects, it has to better evaluate if its projects will make sense for society, for humans."

Elusive Acceptability

Social acceptability is one of the three conditions of sustainable development, along with the environment and economic acceptability. The concept of social acceptability is actually a difficult one to define since it includes social values, concerns for the community and for landscapes, aesthetic choices, other qualitative factors and even in some cases ideas about what's deemed sacred.

Social acceptability is not simply about getting consensus. All the parties involved have to be given the opportunity to talk about their concerns and to influence the direction of a project. In fact, outside extreme emergency situations, Hydro-Québec projects are rarely accepted unanimously. Yet at some point, decisions have to be made, to go ahead with a project or not. That's what it came to in 2002 after public consultations for the case of the high-voltage Hertel-Des Cantons transmission line crossing the Montérégie region. The idea behind the line was to reinforce Hydro-Québec's transmission system after the 1998 ice storm. Opponents to the line got the Quebec Superior Court to declare the government decree illegal and order work to be stopped. But the government, after some wavering, authorized Hydro-Québec to proceed anyway. It judged that in this case, the needs of the collectivity offset the imperative for social acceptability.

Even if electricity consumption in Quebec has stagnated since 2007, Hydro-Québec has constantly had to adjust its transmission system because of population movements. It was actually one of the factors behind the decision to build the new high-voltage transmission line through Saint-Adolphe-d'Howard. Since 2001 the population of the Laurentians region has grown twice as quickly as the rest of Quebec. In fact, the population of Saint-Adolphe-d'Howard grew twice the average of the Laurentians region over the same period: between 2001 and 2011 the village's permanent population grew from 2600 to 3700, a figure that does not include the 6000 cottage owners who live there intermittently. "If we had to supply electricity for a new real estate project or a new ski resort in the area, the local grid would not be able to cope at the moment," explained Marie-Josée, Project Manager, Transmission Lines at Hydro-Québec who coordinates the consultation process Hydro-Québec carries out for each new project.

Hydro-Québec faced the challenge of social acceptability for the first time in 1963 when it was planning the first high-voltage transmission line linking the Manic-Outardes project in the North Shore to Montreal. The original plan was for the line to cross the Saint-Lawrence River at Île d'Orléans, near Quebec City. Hydro-Québec thought the ideal route would be through the quaint village of Sainte-Pétronille. But the village had some influential residents who disagreed with the plan, one of whom was the son of former prime minister Louis St-Laurent. Another

Fig. 14-B: The 735-kV line crossing the St. Lawrence River at Île d'Orléans. Its construction in the 1960s sparked the first real debate about social acceptability. Hydro-Québec originally planned to run the line through the quaint village of Sainte-Pétronille.

was the special advisor to St-Laurent, Louis-Philippe Pigeon—the man who had written the founding law of Hydro-Québec in 1944. Hydro-Québec agreed to run the line across the island but around, not through, Sainte-Pétronille.[2]

Hydro-Québec's projects didn't really start to spark major controversy until the 1970s and 1980s. After years of negotiations, the James Bay Project went ahead. However, the Champigny Project that involved building seven dams on the Jacques Cartier River north of Quebec City, was abandoned. But when Hydro-Québec decided to build a high-voltage line crossing the St. Lawrence River between Grondines and Lotbinière the protest was so intense, Hydro-Québec decided to build an underwater line instead.

Over the last decade, two things about social acceptability have changed. The first is that no scheme is too small to be debated, including 13 km of a 120-kV line. As a result, Hydro-Québec doesn't wait for a project to become polemical before consulting the local populations that will be affected by it; it company does consulting for every new project.

The other change is that social media has made social acceptability more visible to the public, including those not even affected by a project. "Social media has changed the power balance. Citizens can be much more prominent players in the public space than they were before," explains Marie-Josée Gosselin. "Facebook groups pop up spontaneously."

"It's the tyranny of 'now,'" says Sophie Hamel-Dufour. "If an issue is simmering on social media, everybody thinks they have to answer immediately. While for social acceptability, it should be the opposite. People need to take time to get perspective on an issue and reflect on it before doing anything. Quick reactions just end up killing that period of reflection."

Even before the term "social acceptability" became popular, Hydro-Québec had implemented a process for consulting populations that would be affected by its new projects. Whether it's a new substation, a new transmission line or a dam, the process begins with a "general information" session with local elected representatives during which Hydro-Québec presents the project and explains the conditions that justify it. Hydro-Québec's representatives then meet everyone responsible for managing the territory in question, including mayors, municipal general managers and urban planners. These people, in turn, fill Hydro-Québec in on any development projects going on in, or planned for, the area, whether touristic, commercial, residential or other.

Hydro-Québec then uses this information to map the territory. From houses, buildings, trails to upcoming projects, sensitive habitats and landscape, every detail appears on the map. "We want to figure out exactly what's important to a population, from how wide they want a forest clearing to be, to what effect the installation of electrical towers will have on farmers' fields," explains Marie-Josée Gosselin.

Even before the first actual public consultations take place, Hydro-Québec meets the property owners who will be most affected by their project. "We send a special invitation to any resident living within a 600-metre perimeter of the proposed route for the line. Consultations are sometimes a dozen people talking in a church basement. Other times, like in Saint-Adolphe-d'Howard, 200 people show up with posters to denounce the project."

Public consultations almost always lead to modifications in the original project. The high-voltage Chamouchouane-Bout-de-l'île line—a 410-km line with 1000 towers linking Saguenay to Montreal through the Mauricie, Laurentians and Lanaudière regions—is a case in point. The project was widely contested following its announcement in 2015, then, after numerous consultations, modified. For the Attikamek community, Hydro-Québec moved 40 km of the line to avoid passing through a hunting territory. "Their idea was brilliant, and neither more expensive nor more difficult to put into practice than the original plan," says Marie-Josée Gosselin. Further south, the last section of the line was to cross some valuable farmland, and farmers weren't willing

Fig. 14-C: Every new building project, whether for a dam or a line, is preceded by public consultations. At the end of this process, projects rarely go ahead exactly as originally planned. They are usually modified at least once.

to give up a single square metre of field for towers. So Hydro-Québec went back to the drawing board and found a way to use its already existing substations and lines to pass the new line. "That did cost more," recalls Gosselin. "We thought everyone was satisfied, but then we had to manage more protests because we had to clear more land for the new route." In the end, the line passed.

Although it became known as the "Saint-Adolphe line" thanks to the media attention it garnered, the Grand-Brûlé-Dérivation-Saint-Sauveur line (it's actual name) was 41 km long and passed through a number of other villages. While other municipalities affected by the line didn't react strongly, opponents of the 13-km line running through Saint-Adolphe-d'Howard wouldn't give an inch. Hydro-Québec held 95 different consultations for the project. The company studied 11 different scenarios for electric supply, compared four different routes for the line and modified the proposed route many times to place the line further from sensitive habitats and reduce its visual impact. Hydro-Québec even developed a new, shorter, more compact model of electric tower exclusively for the project. "We spent years trying to reach a compromise, and even started the project over from scratch, but the community just wouldn't accept the line," recalls Gosselin.

As Sophie Hamel-Dufour put it, sometimes things just don't work. "When negotiations become too tense it's impossible to arrive at a compromise. There's no more dialogue." There are times when participants are so focused on their problem they lose perspective. In February 2016, the municipality of Saint-Adolphe-d'Howard asked Hydro-Québec to change the path of the line so it would run between Mont Tremblant and the nearby village of Saint-Faustin-Lac-Carré instead—which would have required expropriating house owners in Saint-Faustin, whereas there was no question of expropriation in St-Adolphe-d'Howard.

Not surprisingly, the proposition didn't do much to enhance relations between the villages. Denis Chalifoux, prefect of the regional county municipality (MRC) of the Laurentians told Saint-Adolphe-d'Howard, "Don't leave us holding the bag." The mayor of Saint-Faustin, Pierre Poirier was just as direct. "We were understanding about this issue in the last years, but we have to protect our own citizens. We understand [the concerns of Saint-Adolphe-d'Howard], but you should understand ours too." [3]

That didn't stop protesters in Saint-Adolphe-d'Howard. In 2017, the municipality took its case to the Quebec Superior Court. Hydro-Québec suspended work while the case was being deliberated. In the end, Saint-Adolphe-d'Howard's request for an injunction was rejected. In March 2018, protesters fought back one last time by trying to prevent Hydro-Québec from proceeding with work in the forest, but Hydro-Québec carried on anyway.

To Hide at All Costs

When Julie interviewed Lisette Lapointe, the mayor of Saint-Adolphe-d'Howard was outraged by the fact that, in her view, Hydro-Québec made no effort to preserve rural landscapes. "The North Eastern U.S. states disguise telecommunications towers with fake tree branches. Hydro-Québec defends its way of doing things, but I personally think Quebec deserves better," said Lapointe, who left office in 2017.

The challenge is how, exactly, to disguise power lines, which have much more overall visible impact than telecommunications towers. Even if Hydro-Québec made transmission towers look like Christmas trees, power lines hanging in strips of cleared forests

Fig. 14-D: Roughly 11 percent of Hydro-Québec's distribution grid is buried, compared to less than 1 percent of the high-voltage transmission system. The main reason is that, for reasons of chemistry and physics, it's more difficult to bury high-voltage lines. But it's possible, for a price.

would never be inconspicuous. If Hydro-Québec tried to hide lines behind mountains, the people living on the other side would be stuck seeing them.

The obvious solution is to bury the lines. But the challenge this poses for high-voltage transmission lines (of 44,000 volts or more) is much different than for distribution lines (less than 34,500 volts). Hydro-Québec has buried about 11 percent of its distribution lines; 14,000 km of lines in all, which is roughly the average rate of burial in other Canadian provinces. The decision to bury lines depends mostly on the cost of land and municipal regulations. In the borough of Verdun, all the lines on the Île des Soeurs are buried.

"Managing the impact of burying lines is what makes it expensive in urban areas. In Montreal, just managing traffic while burying lines costs as much as burying the line itself," explains Christian Royer, Engineer , Overhead and Underground Transmission Lines, Hydro-Québec Innovation, Équipement et Services partagés (Department of innovation, equipment and shared services) who designs aerial and underground transmission lines.

Some Quebec municipalities require distribution lines to be buried. For these cases, Hydro-Québec has a program called Embellir les voies publiques (Beautifying roadways) that sets out the rules of how costs will be shared. Municipalities can get real estate developers and homeowners to assume part of the costs.

"Unfortunately, not all municipalities have an urban planner on hand who can work on the regulations and negotiate with Hydro-Québec," says Roger Lanoue. He thinks Quebec's Minister of Municipal Affairs should take the initiative in burying power lines and create a regulatory framework that applies to all municipalities. "The costs of burying hydroelectric lines in that case would be included in Hydro-Québec's distribution costs. Everyone would pay the price, but everyone would benefit equally. And since we have to replace aging distribution lines everything could be buried in a period of 40 years," Lanoue notes. "The discussion over how to make this happen between the Quebec government, who represents municipalities, and Hydro-Québec just never happened," says Lanoue, who recognizes that to some extent, the government is stuck between a rock and a hard place. Burying all lines would cost taxpayers many billions of dollars—300 billion by recent Hydro-Québec estimates. Even a tenth of that could drain all Hydro-Québec's profits for years. Quebeckers might decide they have other priorities.

And that's just for distribution lines. It's an entirely different question for high-voltage transmission lines. The line in Saint-Adolphe-d'Howard falls into this category. Only 1 percent of the 34,500 km of high-voltage lines in Quebec, 225 km in all, are buried and most of these are 12-kV lines in Montreal. There are only two or three short 315-kV lines buried, and again, most of them in Montreal. "We did it because there just wasn't any other place for the lines," says Christian Royer.

Burying high-voltage lines is a very expensive proposition. One reason is that they require a copper conductor instead of the less expensive aluminum conductors used for low-voltage lines. Underground wires also have to be covered with an insulator while wires in the air are just left bare. The first high-voltage lines designed to be buried were invented in Italy. There were insulated with paper soaked in oil, a technique used everywhere until the 1960s, including in Quebec. "At the time we believed underground cables would last about 40 years, but some are still being used today," explains Christian Royer, who estimates that half of the underground network in Quebec still has wires with paper and oil insulation. In the 1970s, manufacturers developed a plastic insulator and Hydro-Québec began using it for its 315-kV high-voltage lines in the 1980s. "The durability has improved, but we don't know exactly how long the lines will last. That adds to the unknown," says Christian Royer.

Everyone in Quebec knows what the "Hygrade sausage syndrome" is. The expression comes from an advertisement that

Fig. 14-E: A "small" 120 kV high-voltage underground cable. It's markedly bigger (weighing 10 kg per metre) and more expensive than an aerial wire but less durable, lasting 40 instead of 80 years.

made a mark on popular culture: "More people eat them because they are fresher. They are fresher because more people eat them." According to Normand Mousseau, physics professor at the Université de Montréal co-chair of the Quebec Commission on Energy Issues in 2013, burying high-voltage power lines in Quebec is the opposite. "Burying lines may be too expensive and those costs might have an effect on electricity rates, but we'll never figure it out if we only bury 10 km of high-voltage lines per year. Since Hydro-Québec isn't even interested in burying lines, it's not actively looking for solutions, so it's not perfecting techniques or figuring out how to get economies of scale."

Éric Martel says it is indeed possible to bury almost all the lines in Hydro-Québec's grid. "Burying lines costs between 40 percent and 300 percent more, depending on the conditions. And it's not always the best option from an environmental perspective. But there is a choice in front of us. Will we be willing to accept rate increases of 3, 4 or 5 percent per year for many years to come just to for aesthetics, without it necessarily doing anything positive for the environment? When I became president, Quebeckers weren't even willing to accept a rate increase of 2.3 percent."

If more cases like Saint-Adolphe-d'Howard pop up in the future, the Quebec government will probably have to step in at some point. There will have to be some serious arbitration done between local interests and those of the majority. "Why should one municipality that values their landscape have the option of burying power lines, while another doesn't?" asks Christian Royer. "But if we decide to bury lines, who will pay for it?"

In negotiations over the export contract with Massachusetts, Hydro-Québec's U.S. partner Eversource said it was prepared to bury 100 km of high-voltage lines. Shouldn't Hydro-Québec and the Quebec government be working on the technology and regulations that will make it possible to bury high-voltage lines in Quebec, too? The answer seems obvious. There are bound to be more cases of Saint-Adolphe-d'Howard and they will come sooner, not later.

Chapter Fifteen

Masters in Their Own House

Among Quebec's jet setting lawyers Paul John Murdoch is in a category of his own. Partly because he actually flies his own plane. Popping here and there in his Piper Seneca PA-34-200T, Murdoch manages to live in the Cree village of Wemindji keeping an office in Montreal and Val-d'Or and making it to regular meetings in Quebec City, Montreal, Chisasibi and Nemaska, in Northern Quebec.

Before he got his pilot's license, Paul John Murdoch was the first member of the Cree Nation to be called to the Quebec Bar, in 2001. Secretary of the Cree Nation Government, he has been the Cree Nation's official representative to the Quebec's National Assembly since 2017 (he is also Cree ambassador to the European Parliament, the United Nations).[1] Hydro-Québec communicates with him several times a week and considers him the Energy Minister of the Grand Council of the Crees.

In 2001, a few months after Paul John Murdoch was admitted to the Quebec Bar, lawyers representing the Grand Council of the Crees pushed him into the negotiations with Bernard Landry's government. The negotiations led to the signature of the *Agreement Respecting a New Relationship Between the Cree Nation and the Government of Quebec*, dubbed "Paix des Braves." This nation-to-nation treaty, signed in February 2002, was meant to

Fig. 15-A: In the winter of 2002, Quebec Premier Bernard Landry and Grand Chief of the Grand Council of the Crees Ted Moses signed a nation-to-nation treaty dubbed "La Paix des Braves." Phase three of the James Bay Project would start.

update the *James Bay and Northern Quebec Agreement* (JBNQA) which had been signed 27 years earlier in 1975 and had been the first treaty between Quebec and the Cree Nations. The two treaties made it possible for Hydro-Québec to develop the potential of the James Bay but they also gave the18,500 Crees in Quebec the possibility of carrying out their own Quiet Revolution and becoming, like Quebeckers, "Masters in their own house."

Among the social acceptability issues Hydro-Québec manages, relations with Indigenous Peoples in Quebec is obviously in a class of its own. Past relations have been contentious and highly mediatized not just in Quebec, but in the United States and Europe. Local and foreign media immediately question the impact on Quebec's Indigenous population every time Hydro-Québec talks about a new export project (see Chapter 9), even if the project doesn't affect their communities.

The media are not wrong in raising the question, but often ignore how things have changed. "Our relationship with Hydro-Québec has never been better than it is today," Paul John Murdoch told us. "We are now in a position to fully participate in the development of the potential of our region. We have demonstrated that we can develop our territory in an acceptable, respectful way that brings improvements. It can be profitable without being destructive."

The Cree, Inuit and Innu communities in Quebec all have close links to Hydro-Québec. It makes sense: fully three quarters of the overall power generated by Hydro-Québec comes from the James Bay and Côte-Nord regions. And Hydro-Québec's activities have to respect some 40 distinct agreements with Indigenous communities. The 15 employees in the department of Relations with Aboriginal communities are Hydro-Québec's ambassadors to the communities, intervening when there's a power outage, when there's work to be done on lines or equipment, and of course, when negotiations happen over new projects.

Indigenous Rights 101

The main thing complicating relations between Hydro-Québec and Quebec's Indigenous communities is that they are largely mediated by different governments. The federal government has the constitutional responsibility to "protect" Indigenous Peoples—it's the spirit and the letter of the *Indian Act*, which does not include Inuit. In principle, the federal government is supposed take care of all aspects of their lives: from health, education and well-being, to policing and justice, even when these are provincial responsibilities. In Quebec, part of these responsibilities, health, education and police, were transferred to the province. In the JBNQA, signed by the Cree Nations and Inuit in 1975, the question of developing the hydroelectric potential of James Bay is just one of *eleven* chapters. The Paix des Braves, signed in 2002, was

Fig. 15-B: The Indians of Canada Pavilion at Expo '67. Quebeckers learned about Indigenous Peoples at Expo just two years before granting them voting rights, but fifteen years after launching the first megaprojects on the Côte-Nord, on Innu territory, and four years before beginning the James Bay Project, without consulting either the Cree or Inuit.

meant to dust off the 1975 settlement, but Hydro-Québec was not even a signatory to the agreement.

Although Quebec and Ottawa are responsible for the major legal, social and economic challenges Indigenous Peoples face, it would be too easy to consider Hydro-Québec a mere bystander in these issues in Quebec. It was, after all, Hydro-Québec projects in Northern Quebec and on the Côte-Nord that woke the different governments up and forced them to negotiate with Indigenous communities in the first place. In 1971, at the time Quebec Premier Robert Bourassa launched what he called the "Project of the Century" to develop James Bay, Indigenous Peoples had only had the right to vote in Quebec for two years—it was nine years after the Canadian government granted them the right to vote. Suffice it to say, there was a lot of catching up to do.

The federal government also had its work cut out for it. Putting aside the fundamentally racist nature of the *Indian Act*, Ottawa has never come up with the necessary funds to actually deliver the protection the law calls for. Nor has it shown the diligence required in order to fulfill these obligations.[2] Indigenous Peoples only obtained the right to hire a lawyer in 1951—nine years before they were granted the right to vote. "Aboriginals didn't have the power to act on the rights that concerned them," sums up Paul John Murdoch. It was only in 1973 that the Supreme Court recognized the existence of Indigenous property rights based on history. As far as Indigenous rights are concerned, Quebeckers must therefore accept the fact that Indigenous Peoples will continue to initiate legal proceedings against Hydro-Québec. After all, it's the nature of the law to evolve through the courts. And governments rarely leave Indigenous communities any other option.

On the question of Indigenous rights, Hydro-Québec started out at about the same spot as the rest of Quebec society: in a state of total ignorance. When Hydro-Québec heard grievances from the Cree for the first time, over the James Bay Project, the company doubled down to find solutions. The result is that today, Hydro-Québec can rightly claim to have contributed to changing mentalities in Quebec. Along with Quebec's provincial police, the Sûreté du Québec, Hydro-Québec was among the first organizations to give its employees awareness training. The department of Relations with Aboriginal communities today trains 100 to 200 employees per year. The workshop Hydro-Québec employees find the most surprising is the one on finance: many employees are surprised to learn that Indigenous Quebeckers pay for electricity just like everybody else.

Between 1971 and 2001, the Quebec government's learning curve was a roller coaster. In 1971, Premier Bourassa announced that the government would be investing billions of dollars to develop the hydroelectric potential of James Bay. Without having consulted the Cree nations first. Naturally, the Cree communities opposed the project and four years later, in 1975, Quebec and Ottawa signed an Agreement with the Cree nations promising to consult them on future projects. But then in 1989, Bourassa went ahead and did it again. He launched the Great Whale project without consulting the Indigenous populations who would be affected by it. Same thing in 1992: the Quebec government failed to consult the Innu community in Mashteuiatsh (near Saint-Félicien in the Saguenay-Lac-Saint-Jean region) about a project for a generating station on the Ashuapmushuan River. There were protests, and the government cancelled the project that year, then cancelled the Great Whale project in 1994. The message seemed to have sunk in: Quebec consulted the Innu people in Sept-Îles and came to an agreement before announcing the dam project on the Sainte-Marguerite River. Then in 1998, Lucien Bouchard and his Newfoundland counterpart Brian Tobin announced a joint project in Gull Island, Labrador, without consulting the Innu people in either province. (Innu protests were successful and the project was abandoned.)

It's hard to understand how something so simple could be so hard for governments to grasp.

Then suddenly, in 1999, relations between Indigenous Peoples and the government of Quebec started improving. That year the government came to an agreement with the Innu in the North Shore community of Pessamit over a project on the Toulnustouc river, a tributary of the Manicouagan River. In 2002, with the Paix des Braves, the government agreed to consult Cree nations on developments being carried out in the region. Two months later, it signed another similar agreement with Inuit. Since then, the same framework has been used in 2004 in Péribonka, in the Saguenay, with the Innu, and in 2008 in La Romaine in the Lower North Shore, also with the Innu. And in 2019, Hydro-Québec will be proceeding this way as it introduces renewable energies into isolated communities.

The Cree Nations

It's surprising how relatively good relations are between the Cree communities, the government of Quebec and Hydro-Québec at the moment, given how badly they started. When the James Bay Project was announced 50 years ago, there was almost no communications between Hydro-Québec and the Cree Nations. The first version of the project called for flooding a large part of Cree territory but no one, either in the government, or from Hydro-Québec, even bothered to let the Cree communities know.

However, Robert Bourassa's "Project of the Century," James Bay, did have an unexpected benefit for the Cree: it united the Cree communities, something that hadn't been seen for a century. Feeling attacked and humiliated, the communities rallied under a political entity, the Grand Council of the Crees. That allowed them to form a single front with a common position. The Cree communities didn't wage war: instead, they fought and won the battle in the courts, armed with impact studies and jurisprudence. The Grand Council has remained in place to this day, despite numerous divisions between "coastal Cree" and the "inland Cree," groups with very distinct customs and dialects.

The Cree Nations got support from a great number of different quarters. First was The Indians of Quebec Association, created in 1965.[3] The Cree also got support from numerous scientists and environmentalists, as well as from the Parti Québécois, which at the time opposed the James Bay Project (see Chapter 2) and was in favour of Quebec developing nuclear energy. And the Cree had a number of lawyers on their side. In November 1973, Justice Albert Malouf issued an injunction to stop work on the James Bay Project. Several days later the Quebec Court of Appeal reversed the decision and work started up again. The opposing judgments

Fig. 15-C: The signature of the Agreement-in-Principle of the JBNQA, in 1974. From the left, federal Minister Jean Chrétien, Chief Billy Diamond, Quebec Minister Gérard-D. Lévesque and Chief Robert Kanatewat.

simply underlined the fact that both parties were right: Indigenous rights exist, but Quebec also has the right to develop its territory. Both camps got the message: "Work it out."

On November 11, 1975, the Cree Nations and Inuit signed the JBNQA. It was the first treaty of modern Quebec and one of the most progressive ever signed over Aboriginal rights, to this day. The agreement promised the Cree $134 million, various other indemnities, the creation of a Cree Regional Authority and a land regime. It was ambitious, but the JBNQA had several shortcomings. While it touched on a wide range of topics, from health and education to policing and justice, many chapters did not define implementation procedures or include dispute-settlement mechanisms. The agreement had to be amended 35 times in the years that followed. Negotiations were difficult and sparked resentment on both sides, but they stumbled on.

By the turn of the millennium, relations between the Cree and the Quebec government had reached an all time low. In 1990, Quebeckers were badly shaken by the Oka Crisis, when Mohawks erected barricades in the suburbs of Montreal—the army was eventually called in to dismantle them. Then the Cree led an audacious campaign against the Great Whale project, forcing the government to abandon its plans and adding another layer to the sense of collective trauma in Quebec. The Cree kayaked down the Hudson River to the New York offices of the United Nations, joined forces with major environmental movements in North America, publicly denounced the damage caused by hydroelectric dams, lobbied U.S leaders and bought an ad in *The New York Times*.[4] Their actions did lasting damage to Quebec's reputation, and their victory left a bitter aftertaste. The end of the Great Whale project and the billions of dollars of lawsuits the Cree brought against the Quebec government had the result of discouraging investment in their region. Grappling with a high birthrate, long-term unemployment and an abysmal dropout rate, the Cree were in urgent need of economic development projects, whether hydroelectric, mining or forestry.

Common sense finally won out in June 2001, but only after a rocky start during a memorable meeting between Grand Chief of the Grand Council of the Crees Ted Moses and Quebec Premier Bernard Landry. Landry had just been named Premier. Ted Moses had decided to make a "courtesy" visit to him to discuss some of the ongoing lawsuits by the Cree. But the meeting soon turned into a shouting match. Each leader vented for 15 minutes without listening to a word the other was saying. And then they just burst out laughing, and finally agreed to start talking. Over the course of the discussion that followed, Landry and Moses came to the conclusion that it would be possible to talk "nation to nation" but only if they wiped the slate clean and forgot the past first. The two leaders spent the next two hours laying the groundwork for three months of top-secret negotiations that would result in a tentative agreement, in October 2001.[5]

After it was ratified by referendum in the nine Cree communities, The Paix des Braves agreement was signed in February 2002. The Cree agreed to drop their lawsuits against the Quebec government, and manage their own affairs. The Quebec government agreed to pay them $3.5 billion in royalties over the next 50 years to exploit hydroelectricity, mines and forestry.[6] "In a few months we managed to settle a series of old lawsuits over policing, justice, education and health," recalls Paul John Murdoch, who accompanied Ted Moses to the public assemblies that were held in each community prior to the referendum. "The Paix des Braves is a gamble on the future, on being able to manage things ourselves and take responsibility for our affairs. It was very revolutionary."

Responsibility for the hydroelectricity dossier fell on Paul John Murdoch's shoulders during the negotiations. Murdoch was at the forefront of negotiations over the Eastmain-Rupert Rivers project, and over implementing all mitigation measures related to the social and environmental impact of the project. "Ted had to convince 15,000 Cree, but I was responsible for dealing with 22,000 hostile Hydro-Québec employees. There was a lot of antipathy at the early meeting. The fact that the words 'without prejudice' weren't written on the agenda made everyone nervous."

The Cree communities were involved in all the planning related to the Eastmain-Rupert project, not just regarding mitigation measures but also decisions about planning and even the design of the reservoir. "There were three proposed routes for diverting the Rupert River, but they all involved flooding the Champion Lake. That lake was important to the population of Nemaska. So a Cree trapper proposed that instead of flooding a large territory, they divide the reservoir of the Rupert River and build a 3-km long control tunnel to divert water under the lake from one reservoir to the next. The solution had the effect of reducing overall compensation to the Cree, and reducing the surface of the reservoir by 30 percent," says Paul John Murdoch. "But it allowed the Nemaska people to preserve the lake. We had meetings like that every week, over every subject."

In the negotiations following the Paix des Braves, the Grand Council of the Crees and Hydro-Québec agreed that the Cree would get about 10 percent of the estimated 5500 construction jobs created by the Eastmain-Rupert Project, as well as 150 permanent jobs at Hydro-Québec. Paul John Murdoch consolidated all the activities related to environmental assessments and training into one organization, the Niskamoon Corporation. "At first we trained our people in Ontario and tried to teach them French but then we realized that didn't make sense. So we created a training centre in Rouyn-Noranda instead where the training would be offered only in French. We could do this because the decision was up to us. If the government had tried to force our hand, it would never have worked."

Power system electrician Rebecca Diamond, one of the Cree who trained at the Centre Polymétier de Rouyn-Noranda, was part of the last cohort of electricians to take advantage of Hydro-Québec's agreement to hire 150 Cree employees. When Jean-Benoît met her at the Eastmain-1 generating station, the 42 year old was finishing an eight-day work period before returning to spend six days with her family in Abitibi. Having grown up and gone to school in French in the town La Sarre in the Abitibi region, Diamond had not needed French lessons. However to be accepted for electro-technical training program, she did need to finish secondary school. She went back to school at age 32. Today she is one of 10 Cree among the 71 employees of the Eastmain-1 and 1-A generating stations. "I wanted to prove that a native woman could do it," she told Jean-Benoît. "I was a little concerned about being accepted, but everything went well. I'm very proud. And so are my kids."

Rubens Durocher, Manager of Eastmain 1 and 1-A generating stations, says Rebecca Diamond probably had an easier time than her predecessors getting work because the program of reserved jobs at Hydro-Québec had been running for a while before she arrived. "At the beginning, things were tense with Hydro-Québec's union. Some employees who didn't have permanent jobs yet were training Aboriginal employees who got permanent status right after the probationary period. There was a lot of explaining to do. Niskamoon also had to update its training techniques," says Rubens Durocher. There are now a 100 Cree among the 900 employees in James Bay. But the program has ended. "In the future, Aboriginal candidates will have to follow the regular channel."

Inuit

According to the framework set out in the JBNQA, Hydro-Québec must use diplomacy in its dealings with Indigenous communities even when performing an activity that seems to be beneficial to them. For example, replacing part of the diesel used by power generators in the communities of Quebec's 12,000 Inuit with alternative energy sources sounds like a win-win solution for Hydro-Québec and Inuit—an obvious solution to a clear problem. The law requires Hydro-Québec to supply electricity to all Quebeckers at the same price, including the 35,000 residents of the 25 off-grid systems that supply 30 isolated communities, almost all of which run on diesel. Alternative energies would allow Hydro-Québec to cut costs and reduce GHG emissions in one shot. Who could possibly be against that?

The problem is that Nunavik is not the South, so things don't necessary work according to the same logic. The question is: who sells the diesel that supplies the Hydro-Québec's generators? The answer: Inuit themselves. Who owns the land where solar panels or wind mills will be installed? The land corporations of Inuit villages, which, like Makivik and Kativik, were born of the JBNQA. What are the main businesses operating in the Inuit communities? There are fourteen cooperatives in the Inuit villages gathered

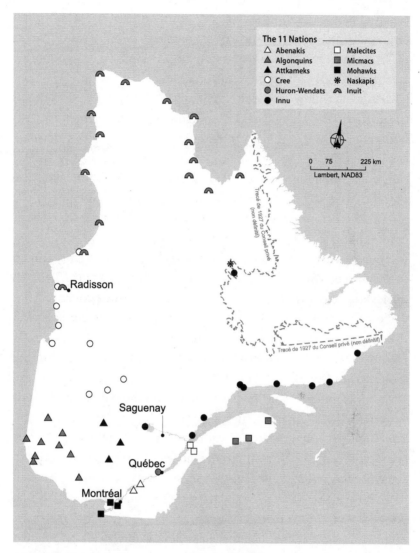

Fig 15-D The 11 Indigenous nations in Quebec. James Bay, on Cree and Inuit territory, supplies 20,000 megawatts of power, or half of Hydro-Québec's installed capacity. The Côte-Nord, Innu territory, supplies another third, about 13,000 megawatts.

together under the umbrella of the Fédération des cooperatives du Nouveau-Québec (Federation of cooperatives of Nouveau-Québec, FCNQ). And then there is the Makivik Corporation, the economic arm of the Kativik Regional Government.

In short, whatever the motive—whether it's the well-being of the planet, or Hydro-Québec's bottom line, or a healthier, less noisy local environment—replacing diesel generators or solar generators poses a real problem for Inuit, who risk losing tens of million of dollars in gas sales. Even when Hydro-Québec has the power to manage its generator however it wants, the company has to take the economic and political realities of Inuit into account before doing so.

Inuit, meanwhile, have realized they'll have to get organized in order to be able to benefit from the energy transition. "The two principal players are Makivik and the FCNQ, and neither are what you'd call small operations," says Vincent Desormeaux, Advisor, —Energy Transition and Indigenous Affairs, Off-grid Systems for Hydro-Québec. Makavik, which is responsible for managing the money from the JBNQA and different agreements, made the wise decision to invest in its own airline company, Air Inuit. Then it bought First Air airline, which serves the entire Canadian North. The FCNQ, meanwhile, took over the contract Shell had on its territory and consolidated all purchases (everything from energy to eggs) of the 14 local cooperatives, a total of $450 million per year.

In February 2017, when Hydro-Québec had just launched its pilot project to install solar panels in the small Northern community of Quaqtaq (see Chapter 6), Makivik and the FCNQ signed an agreement to create Tarquti Energy, a joint venture to develop renewable energy in Nunavik.[7] "Their first goal is to become a supplier of renewable energies to Hydro-Québec," says Desormeaux. "In the long run, they want to offer the same services to mining companies, prospectors and outfitters."

The negotiations are taking a while because Tarquti wants to earn money, as do the cooperatives and land corporations operating in villages. Hydro-Québec's aim is to reduce costs of producing electricity in the North. Vincent Desormeaux sees the advantages of a partnership between Hydro-Québec and Tarquti. For starters, Tarquti is in a better position to negotiate with local communities than Hydro-Québec is. Working through Tarquti will also give Hydro-Québec access to subsidies from the federal government that cover up to 75 percent of the costs, as opposed to 50 percent. "All energy projects in the North are subsidized, without exception," says Desormeaux. "The pilot project we are carrying out in Quaqtaq isn't just a technical test. We are using it to create a political-financial model we can use to figure out if the energy transition is a viable option for all off-grid systems. It's not as easy as it sounds."

The Innu Nation

While the Cree and Inuit can work with Hydro-Québec through comprehensive treaties, the Innu—once known as the "Montagnais"—have never had anything like this. This is quite an amazing given that more than a quarter of Hydro-Québec's power comes from the Côte-Nord, their territory: an immense swath of land stretching east from the Saguenay Fjord to Blanc-Sablon on the border of Labrador, then north to Schefferville (200 km north of Labrador City).

In 1953, when Hydro-Québec began work on the Bersimis generating station, the Innu people had only obtained the right to hire a lawyer two years earlier. They would have to wait another 16 years to be able to vote in a Quebec election in 1969, nine years after gaining the right to vote in federal elections. But the Innu people were not only deprived of these basic democratic rights, they also lacked the political or cultural organizations that would make it possible to intervene in public affairs. When the Indian Association of Quebec came into being, in 1965, the Manicouagan-Outardes project was already producing its first megawatts. So the Innu people never had any bargaining power to force Hydro-Québec to negotiate.[8]

The 20,000 Innu in Quebec are the largest Indigenous nation in the province. Contrary to Cree and Inuit communities, who were traditionally isolated from Quebec society, the Innu people have had sustained contact with Quebec society. A full 40 percent of Innu live outside reserves. "Hundreds of Innu have worked on Hydro-Québec's big work sites," says Martin Dufour, Chief of the Conseil de la Première Nation des Innus Essipit (Council of Innu First Nations, in Essipit), near Les Escoumins on the North Shore of the St. Lawrence River.

While the majority of Cree speak Cree and English, the Innu people speak Innu and French. The nation is divided between those in the west who have close relations with the Quebec population , and those in the east who have much less contact with mainstream Quebec society. In the middle of this spread there is an urban group in Sept-îles. Almost all the communities are coastal, except three groups, who are also distinguished by geography: there is an Innu community in Mashteuiatsh, on the eastern shore of Lac Saint-Jean, and one other near Schefferville, Matimekosh. Nearby is the community of Kawawachikamach, home to the Naskapis.

The Innu Nation missed a historical opportunity in 1975 during the negotiations leading to the JBNQA agreement. The Grand Chief of Grand Council of the Crees, Billy Diamond, invited the Innu Nation to participate in the negotiations, but they didn't join in. Only the Naskapi, near Schefferville, negotiated a separate agreement: the Northeastern Quebec Agreement, ratified in 1978.

The Innu people established a Conseil des Attikameks et des Montagnais (Council of Attikameks and Montagnais, CAM) in 1979 to negotiate with the Quebec government over a territorial land claim agreement. But the CAM splintered in 1994. In 2004, the Innu of Mashteuiatsh, Essipit, Pessamit and Nutashkuan reached the stage of an Agreement-in-principle of a general nature

("Entente de principe d'ordre general"), but nothing has ever been signed.[9] "We're not in a rush," says Martin Dufour. "The Minister of Indigenous and Northern Affairs is being completely revamped right now and we want to see what will come of that."

Until the Innu Nation, Quebec and Ottawa agree on a treaty that will create a clear frame for Hydro-Québec's actions, things will carry on the way they always have. Each project will be negotiated separately, each time starting from scratch. In the case of the La Romaine Complex, Hydro-Québec was able to sign three agreements with the four communities involved (Unamen Shipu, Nutashkuan, Ekuantshit and Pakuashipi). These agreements ensure payment, guarantees of jobs and contracts with local businesses, as well as a series of environmental measures. But arriving at the agreements was a turbulent process, marked by blockades and numerous demands for amendments. And there is another factor that ruptures the continuity of discussions and hinders coordination between interlocutors: the leaders of each community are up for election every two, three or four years.

Given how stormy negotiations were for the La Romaine project, a weather warning is in effect for the Apuiat wind farm project in Port-Cartier. The project for 57 windmills with a capacity of 200 megawatts will be carried out jointly by the renewable energy company Boralex and the nine Innu communities of the region. A lot of ink was spilled over this project before and after Quebec's provincial elections in 2018. The peculiarity of the Apuiat project was that, for the first time since 1994, it united the Innu Nation.[10] "Right now we have formed an informal council of nations," explains Martin Dufour. "We realized we weren't being treated the same way as the Cree or Inuit. We should have been talking long ago but we are, finally."

Hydro-Québec's managers had serious reservations about the Apuiat project from the start. The Liberal government seemed determined to go through with the contract even though Hydro-Québec was already grappling with an energy surplus, mostly for political reasons. Pierre Moreau, Quebec's Minister of Energy and Natural Resources at the time, went as far as threatening to fire Hydro-Québec President and CEO Éric Martel if he didn't comply.[11] In the end, Quebec's new business-friendly CAQ government pulled the plug on the Apuiat project. "We are disappointed, but not in shock. We understand the problem of energy surplus," says Martin Dufour. "But the government promised that the next wind farm project would be ours."

The Innu people will definitely have a stronger hand in future negotiations over wind energy projects or any energy project for that matter. Hydro-Québec has said for years that it would wait until it finished building La Romaine before deciding on any more dams. However the company is changing its tune and looking more favourably on the new premier's export projects. It also has to consider the end of the Churchill Falls contract in 2041. "From the draft proposal, to a finished dam project, it takes 18 to 20 years," says Éric Martel. "So if we want a finished project by 2039–2040, we have to have chosen the river and launched the project by 2021 or 2022."

Officially, Hydro-Québec hasn't named any specific rivers, but a few ideas have been circulating over the decades. For instance, Hydro-Québec referred to a possible project including Petit Mécatina (just east of La Romaine) and others on Inuit territory including Nastapoka, Caniapiscau, the George River and Great Whale River in 2004.[12] Five years later, as part of the Plan Nord, Premier Jean Charest added the Caniapiscau Reservoir basin, as well as the rivière-aux-Feuilles in the Ungava Peninsula. The most probable option is the Petit Mécatina River in Quebec's North Shore region,[13] a 1200-megawatt project. The power line leaving the four generating stations of La Romaine was calibrated for additional generation from the Petit Mécatina.

The other likely project is Gull Island, Labrador, which sits on land claimed by the Innu people. If it goes forward, the Innu community will gain the bargaining power that has eluded them for so long. The risk of thunder and lightening will be high in Quebec's North Shore in the coming years, and when this happens, Hydro-Québec will likely serve as the Quebec government's lightening rod.

Chapter Sixteen

Political Highs and Lows

No power system in the world is immune to bad political decisions, not even Hydro-Québec's. The governments of Ontario and Newfoundland both played a role in the respective setbacks of their provincial systems. In Newfoundland, Premier Danny Williams decided to start the Muskrat Falls generating station project in 2010. It was based on overly optimistic hypotheses, ended up costing three times more than projections and tripled electricity rates in the province. In Ontario, the choice and then mismanagement of nuclear energy cost customers and taxpayers tens of billions of dollars.

As Ontario Hydro's brush with bankruptcy in 1999 showed, no public power system is "too big to fail." In Quebec, Hydro-Québec's main financial and technical difficulty at the moment is the 41 wind-power supply contracts the government forced on the company in the mid-2000s. They aren't a threat to the viability of the system, but the contracts are costing Hydro-Québec hundreds of millions of dollars per year when it is already struggling with an energy surplus. It's not easy to know where to lay the blame for Hydro-Québec's bad decisions: on the company's managers or its single shareholder (the Quebec government)? Two cases in point: the strange contract for the Bécancour natural gas-fired generating station, and the premature liquidation of Hydro-Québec

International in 2006. As the proverb says, success has many fathers, but failure is an orphan.

The question of "who decides" will become all the more important as Hydro-Québec takes on the ambitious program to expand beyond its traditional markets. The company will require constant support from both present and future governments. President and CEO Éric Martel is getting ready to invest billions of dollars in developing new markets, through exporting electricity, electrifying transportation, commercializing Hydro-Québec's patents and acquiring foreign infrastructure. The Quebec government has not only agreed to pursue these goals, the government was involved in formulating them in the first place. At the same time, these investments will take many years to bear fruit, so Hydro-Québec's single shareholder will have to be patiently supportive through changes of governments, ministers and deputy ministers. And the government will have to be ready to accept the inevitable criticism if things do not go exactly as planned.

In other words, Hydro-Québec needs a government that is not only ready to dive in, but will have the backbone to support it over the long term.

It's remarkable that Hydro-Québec has managed to weather political uncertainties over the decades. In the 75 years it has existed, confidence in the government-owned company has always prevailed in the face of partisan politics. Hydro-Québec was created by a Liberal government in 1944, but faced a new government within months: it had to live with the Union nationale until the Liberals returned to power in 1960. Over 75 years, Hydro-Québec has had to deal with 12 different governments and 29 energy ministers.

While Hydro-Québec makes its decisions about what, where and how to build over a time span of 25 or 50 years, one year is an eternity in politics. Elected politicians are notoriously sensitive to shifts in public opinion. That's why a debate about whether to partially or totally privatize Hydro-Québec re-emerges every five

Ministers responsible for Hydro-Québec

Order	Minister	Party	Date of appointment
1	John Samuel Bourque	UN	July 21, 1945
2	Daniel Johnson	UN	April 30, 1958
3	René Lévesque	Lib	July 5, 1960
4	Gaston Binette	Lib	January 19, 1966
5	Daniel Johnson	UN	June 16, 1966
6	Paul-Émile Allard	UN	October 31, 1967
7	Jean-Gilles Massé	Lib	May 12, 1970
8	Jean Cournoyer	Lib	July 30, 1975
9	Yves Bérubé	PQ	November 26, 1976
10	Yves Duhaime	PQ	April 30, 1981
11	Jean-Guy Rodrigue	PQ	November 27, 1984
12	Michel Clair	PQ	October 16, 1985
13	John Ciaccia	Lib	December 12, 1985
14	Lise Bacon	Lib	October 11, 1989
15	Christos Sirros	Lib	January 11, 1994
16	François Gendron	PQ	September 16, 1994
17	Guy Chevrette	PQ	January 29, 1996
18	Jacques Brassard	PQ	December 15, 1998
19	Gilles Baril	PQ	January 30, 2002
20	François Gendron	PQ	February 13, 2002
21	Sam Hamad	Lib	April 29, 2003
22	Pierre Corbeil	Lib	February 18, 2005
23	Claude Béchard	Lib	April 18, 2007
24	Nathalie Normandeau	Lib	June 23, 2009
25	Clément Gignac	Lib	September 7, 2011
26	Martine Ouellet	PQ	September 10, 2012
27	Pierre Arcand	Lib	October 23, 2014
28	Pierre Moreau	Lib	October 11, 2017
29	Jonatan Julien	CAQ	October 18, 2018

Fig. 16-A: Hydro-Québec has seen 29 energy ministers, though the actual name of the ministry ("Energy and Natural Resources," since 2014) has changed eight times in 75 years.

or ten years. (Though the issue is quickly shelved since Hydro-Québec has always been profitable.)

Then there are those who try to manipulate public opinion to influence political decisions. Since the mid-1990s, a powerful lobby has risen across the planet, and in Quebec, to promote "renewable energy," understood as solar and wind power. This is not, of course, a bad thing. Yet university professor Pierre-Olivier Pineau, Research Chair in Energy Sector Management at Montreal's HEC business school, is struck by how underrepresented hydroelectricity is in the discussion of renewable energy sources. "The problem is not unique to Quebec. It's continental. Hydroelectricity is the stuff of large public utilities who exist, but don't lobby. Private solar and wind energy producers don't hesitate a second in hiring lobbyists to promote their energy. It's gotten to the point where hydroelectricity has no voice in influencing public opinion." The problem resurfaces every time Hydro-Québec raises questions about wind power contracts (like in Apuiat) or makes a move toward opening new markets outside of Quebec (like the recent contract with Massachusetts). Few people instinctively think of hydroelectricity as a renewable source of energy the same way they do wind or solar energy.

Hydro-Québec has never been independent from the government and it never will be. According to André Bolduc, former economist and author of many historical books on Hydro-Québec, this lack of independence has its drawbacks. "Back when Hydro-Québec was an energy commission, its managers could hold their own with the government. They weren't afraid to confront the government and often won their case. When Hydro-Québec was commercialized at the beginning of the 1980s, it lost its freedom of speech vis-à-vis the government," he says. "In the matter of wind power, no one seems to have examined which energy source would be optimal. Hydro-Québec has never been able to make itself heard on the question. It doesn't have the freedom to contribute to the debate that a company of its size should have." This lack of independence was what made the controversy over the Apuiat project so unusual: it's been a long time since any Hydro-Québec CEO has openly defied a minister.[1] The government went as far as threatening to fire Éric Martel, but in the end the president of Hydro-Québec's Board of Directors, Michael J. Penner, took the fall and resigned a few weeks after the election of François Legault's CAQ government. (He was replaced by Jacynthe Côté, former CEO of Rio Tinto Alcan.)

Hydro-Québec will never enjoy the kind of independence from the government that Quebec's Caisse de depot et placement has. The Caisse has an economic development mandate, but its primary mission is just to make its depositors' money grow. So the decisions of the Caisse are necessarily conservative. A financial behemoth of $300 billion, it manages a very large proportion of Quebec's pension funds, starting with those of civil servants, who are not willing to risk their returns investing in far-fetched government schemes.

Hydro-Québec's decisions are by definition political. It makes visionary, at times audacious decisions, but theses have the potential to immediately affect the lives and well-being of all Quebeckers. Moreover, because Hydro-Québec is so profitable the government expects it to contribute to public finance through the dividends it pays to the Quebec Treasury. And the government also expects Hydro-Québec to makes decisions in accordance with Quebec's energy policy and support regional economic development.

What's more, every Quebec government faces an opposition that never misses the opportunity to question Hydro-Québec's activities—whether it's over the relevance of a new dam project or the proper management of a power outage at the end of some country road. The Minister responsible for energy has to be constantly informed of Hydro-Québec's activities. And other technical ministers (Finance, Treasury, Environment, Commerce, Economic Development) also get their say in energy matters. Then there's the premier, whose office, until 2003, was actually located in Hydro-Québec's headquarters.

Shooting Itself in the Foot

Hydro-Québec would be subject to a high degree of scrutiny even if it were not owned by the government. Electricity is of such huge economic, strategic and symbolic importance that governments must always regulate the actions and orient the choices of companies. Like all companies, Hydro-Québec makes some bad calls and occasionally shoots itself in the foot. But few businesses have the strategic and economic value of electricity. Who could get by without electricity?

Hydro-Québec didn't have the government's help getting stuck in the 20-year saga developing electric cars. And the government didn't help it sabotage the first version of Hydro-Québec International. At various times in the past, the company has operated blindly. Between 1996 and 2015 Hydro-Québec hired two CEOs in a row from the natural gas industry, André Caillé and Thierry Vandal, who then went on to surround themselves with managers from the same sector. Someone in their entourage really should have seen the shale gas phenomenon coming before it was too late: the competition from this sector did serious long-term damage to Hydro-Québec's export prospects.

Hydro has also had problems with its "clans," not political, but technical. This is part of the reason it took seven years just *deciding* to develop its 735-kV transmission lines—seven whole years. (It only took half that time to actually build the lines and get them operating.) The reason for this slow decision-making was that historically, a large group of employees considered nuclear power a viable option, favouring it over even hydroelectricity. Another "clan" believed the same thing about wind power. The debates between different groups aren't always in good faith: there are winners and losers. Some parties fight like tigers, convinced they are defending the common good of Quebec (even when they are driven by self-interest). And the winners in these clashes aren't always the ones defending the best option.

One of the great internal debates at Hydro-Québec at the moment is whether to start a new dam project after the La Romaine Complex is completed in 2020. There is another debate over how to react to the soaring popularity of solar panels. Will solar turn out to be a passing trend or a lasting change that revolutionizes the electricity market in Quebec? Will solar self-generation give Hydro-Québec the chance to put off investing in a new dam or transmission lines?

And what about Hydro-Québec's energy surplus? Some argue for using the surplus to develop new export markets. Others think new markets should come from inside Quebec. The problem is, there's nothing eternal about this "overabundance." The surplus actually could be erased by three or four years of drier than normal weather. A new aluminum smelter in Quebec, a couple of new export contracts or three or four big data centres could also wipe it out. And if any of those things happened, or all of them, what would Hydro-Québec's margin of maneuver really be?

The same debates are taking place at Hydro-Québec over the questions of storage batteries, advanced home automation technology, the electrification of transportation, commercializing patents and acquiring foreign electricity infrastructure.

As Quebeckers' perception and use of energy changes over time, Hydro-Québec also has to stay on high alert. But again, what's the line between being "ready" for these changes and doing too much, or not doing enough to prepare for them? No matter what it does, Hydro-Québec will always run the risk of taking the wrong path and ending up at a dead end.

But Hydro-Québec cannot afford to sit on its hands and wait for trends to emerge.

The government, meanwhile, must keep a constant watch over what is arguably its greatest single asset. Quebec has put in place a series of countermeasures to prevent any single person from making a crazy decision that jeopardizes what it has built. One of these is the Régie de l'énergie, created in 1996 to depoliticize electricity by regulating rates and services for the transmission and distribution of electricity and natural gas. But the issues of electricity generation and the government's energy policy are not up to the Régie de l'énergie's to decide. So from time to time, when the Régie fails to take Quebec's energy policy or the struggle against climate change into account in its decisions, the government has stepped in to reorient its actions.

The Régie de l'énergie is not the only counterbalance to political influence over the energy market. The power system's very cohesiveness helps do this. It's unlikely any government in Quebec would ever be able to completely break up the province's integrated hydroelectric grid. By contrast, there is no entity (government-run or otherwise) in say, Ontario, or Massachusetts, or the State of New York, that watches over the entire power system. In the 1990s, Ontario Hydro had to contend with 300 separate municipal grids that distributed its electricity. These have been reduced to 76 players today, but Ontario Hydro has been sliced up into five distinct enterprises, in the meantime. This division of territory means there has to be constant political intervention to mediate between competing interests. That increases the risk politicians could make ill-informed or partisan-inspired choices that do lasting damage to the system. For example, in 2003 the Liberal government of Dalton McGuinty wanted to create a "green economy" as quickly as possible, so it set out to turn the province into a pioneer in renewable energies. McGuinty's government decided to do this by encouraging the solar and wind power sectors, mainly through extremely lucrative supply contracts (Ontarians pay much more than Quebeckers do for wind power). But in doing so, the government not only burdened its electricity industry with high energy prices, it practically broke the system by introducing almost unmanageable problems.[2]

This same thing could happen in Quebec, but it's less likely to for the simple reason that Hydro-Québec is always thinking of the system as a whole. Even if a minister threatens to fire a CEO who refuses to follow government orders, the CEO of Hydro-Québec will have a strong voice in decision-making if he (or she) has the backbone to use it.

And even if Hydro-Québec found itself in the hands of a complacent or ineffective government, the company's CEO still couldn't do whatever he or she wanted to with Quebec's power monopoly. That's because the company is subject to stricter rules of accountability than those any ministry is required to adhere to. The public and the parliamentary opposition have the tools they need to put Hydro-Québec back on the right path if it ever wanders. In 2003-2004, when Hydro-Québec wanted to launch the Suroît project to build an 885-megawatt gas-fired generating station near Beauharnois, the announcement sparked such a strong public backlash that Hydro-Québec's executives quickly backtracked and cancelled it.

2044 is Around the Corner

One can lose perspective by concentrating on the negative. Although Hydro-Québec is sure to hit some roadblocks along the way, the overall picture at the moment is positive.

Quebeckers are right to be proud of their government-owned power system. Hydro-Québec is one of the largest, one of the most profitable power systems on the continent, and on the planet—and definitely the "greenest," producing electricity from 99.8 percent renewable sources. Quebec has been both an early and visionary player in the transition to renewable energy sources, having started in the 1970s and 1980s, or two generations before the rest of the continent, much earlier than California, Ontario or Vermont. Quebeckers collectively accepted the costs of this transition long before the expression "energy transition" was even coined—owing largely to the fact that Quebec also managed to keep its energy prices lower than anywhere else on the continent.

Hydro-Québec can't predict the future. It has made mistakes and a few puzzling decisions. And yes, the government has gotten mixed up in Hydro-Québec's business more than once. Yet in over 75 years of operation, none of these factors spelled the death of Hydro-Québec or overall put its customers at risk.

The next 25 years will bring exciting changes with unknown consequences. There are new technologies like solar panels, mega-batteries used for storing energy, and advanced home automation technology on the horizon. Quebec also has promising new markets, like electrifying transportation or decarbonizing the economies of its neighbouring states or provinces. New energy consumption habits and attitudes are already taking shape, whether it's the swing toward distributed generation or energy use that is more respectful toward the environment. These challenges could look like obstacles to Hydro-Québec. Instead, Hydro-Québec is treating them as opportunities to help the company move forward.

If the past is any indication of what lies ahead, Hydro-Québec's future will be electrifying.

Notes

Introduction. The Future of Electricity

1. Charles Lecavalier, "Hydro-Québec face à une 'spirale de la mort'," *Le Journal de Québec*, January 9, 2018.
2. Jean-Benoît Nadeau and Julie Barlow, "Les sept travaux d'Hydro-Québec," *L'actualité*, March 6, 2019.

Chapter One. A Tour de Force

1. Richard Rhodes, *Energy, A Human History*, (New York: Simon & Schuster, 2018), 171.
2. For more details, see Appendix 1, p. 297.

Chapter Two. Technical Prowess

1. The acronym IREQ is based on the institute's original name: Institut de recherche en électricité du Québec.
2. Alain Chanlat, *Gestion et culture d'entreprise : le cheminement d'Hydro-Québec*, with the collaboration of André Bolduc and Daniel Larouche, 56, Montréal, Québec Amérique, 1984, 250 pages.
3. Bersimis is another name for the Betsiamites River, but is no longer used.
4. Construction of the original project lasted from 1958 to 1978. In 1989, another generating station, Manic-5-PA, was added, with 1064 megawatts. Then in 2005 the Toulnustouc generating station, on the river of the same name, went into service, supplying 526 more megawatts of power. The Manic-Outardes complex now has nine generating stations supplying a total of 8300 megawatts of power.
5. Jean-Benoît Nadeau, "Une histoire électrisante," *Québec Science*, November, 2012.
6. Engineers like giving themselves some margin of maneuver. A 315,000-volt line is actually a 300,000-volt line with 5 percent additional capacity, which

makes a total of 315,000. A 120,000-volt line is actually a line with a 115,000 capacity, to which 5 percent is adding, making a total of 120,000. A 735,000-volt line is actually a 700,000-volt line, increased by 5 percent These are approximations: technicians can add more voltage to a line by more than a 5 percent margin, but only of absolute necessity, because it damages the line.

7.   Jean-Louis Fleury, *Les coureurs de lignes : histoire du transport de l'électricité au Québec*, (Montreal: Stanké), 113-114.

8.   As Hydro-Québec engineers would discover years later, 735-kV lines had so much capacity that only one line may be required for smaller projects. In these cases, two lines are installed so the system doesn't rely on a single line.

9.   Fleury, *Les coureurs*, 226.

10.   Ibid., 226.

11.   The Soviets were the only ones experimenting with higher voltages in the 1970s. For several years, starting in 1985, they operated at a stunning 1.15-million-volt (1150kV) line that was 432 km long, in Kazakhstan. This high-voltage network was never developed further, and since 1991, has only run at half capacity.

12.   Guy Deshaies, "Prématuré pour l'Hydro, dépassé pour Parizeau!," *Le Devoir*, May 1, 1971.

Chapter Three. The Electric Revolution

1.   Jacques Beauchamp, "La naissance d'Hydro-Québec racontée par Stéphane Savard," Radio-Canada's *Aujourd'hui l'histoire*, January 22, 2018.

2.   First annual report of the Commission hydroélectrique de Québec, 1944.

3.   Marcel Labelle, *Adélard Godbout, précurseur de la Révolution tranquille*, (Montreal: Lidec), 331.

4.   Jean-Benoît Nadeau, "Une histoire électrisante," *Québec Science*, November, 2012.

5.   The government never wanted to nationalize the 30 or so industries in Quebec that produce electricity for their own use, the biggest of which was Alcan, based in the region of Saguenay & Lac-Saint-Jean. The aluminum producer drew 2000 megawatts of power, or 500 MW more than the Shawinigan Water & Power Company. Even today, the electricity department of Rio Tinto Alcan has 640 employees who operate 28 dams and 6 generating stations that produce 2042 megawatts of electricity. It remains the second largest producer of hydroelectricity in Quebec.

6.   Matthew McClearn, "Ontario kept billions of borrowed money off its balance sheet. Here's how," *The Globe and Mail*, April 21, 2018.

Chapter Four. The Future Is Here

1. For more details, see Appendix 1, p. 297.
2. Hydro-Québec has two thermal power stations in Bécancour. The first is a 411-watt gas turbine power plant put in service in 1993 that operated as an auxiliary to the Gentilly-2 nuclear power plant. After that station was decommissioned, Hydro-Québec kept the gas turbine plant for emergency use. The second thermal station, property of TransCanada Energy, was built in the mid-2000s to offset Hydro-Québec's short-term power shortfalls. It operated in 2007–2008 but not since, because electricity demand stagnated.
3. For the anecdote, Shawinigan Engineering, a subsidiary of Shawinigan Power became shareholder of the British Newfoundland Corporation (Brinco) in 1958 to develop the project. Hydro-Québec became a partner after the nationalization of Shawinigan Power in 1963.
4. Pierre-Olivier Pineau, "L'intégration des secteurs de l'électricité au Canada," *L'Idée fédérale*, March 2012, 15, note 4.
5. Nadeau and Barlow, "Les sept travaux d'Hydro-Québec," *L'actualité*, March 6, 2019.

Chapter Six. Testing the Power System

1. Diane Cardwell, "Solar Power Battle Puts Hawaii at Forefront of Worldwide Changes," *The New York Times*, April 18, 2015.
2. Andrew Burger, "Project Spotlight: The Philadelphia Navy Yard Microgrid," *microgridknowledge.com*, August 16, 2017.
3. Roger Andrews, "The California Duck Curve isn't confined to California," *Energy Matters*, November 15, 2017.
4. Justin Worland, "After years of torrid growth, residential solar power faces serious growing pains," *Time*, April 4, 2018.

Chapter Seven. The Smart Bill is Coming

1. Pierre Couture, "L'électricité la moins chère en Amérique du Nord est au Québec," *Le Journal de Québec*, October 13, 2018.
2. The two programs came about as the result of a pilot project called *Heure juste* carried out with 2200 customers in the cities of Trois-Rivières, Saint-Jean-sur-Richelieu, Sept-Îles and Val-d'Or in 2010. The program didn't produce the expected results because the rates proposed weren't attractive enough

Chapter Eight. Creating Demand

1. Antoine Jacob, "La Norvège, championne de la voiture électrique," *LesEchos*, March 30, 2017.

Chapter Nine. The Battery of the North East

1. Michael Overturf, "How the Price for Power Is Set," *Forbes*, December 26, 2012.

2. In May 2019 the mayor of New York City announced he would be buying large quantities of hydroelectricity from Hydro-Québec, as much as Massachusetts, and said he hoped negotiations would conclude by 2020. At the time of printing, Hydro-Québec was even considering paying for the Champlain Hudson Power Express, a high-voltage underwater power cable between Montreal and New York.

3. François Normand, "Le plan de match d'Hydro-Québec pour exporter son électricité d'ici 2030," *Les Affaires*, September 16, 2017.

4. Adrian Morrow and Bertrand Marotte, "Ontario to buy hydroelectricity from Quebec," *The Globe and Mail*, October 21, 2016.

5. Konrad Yakabuski, "Time for Ontario and Quebec to talk hydro," *The Globe and Mail*, October 24, 2018.

6. Pierre-Olivier Pineau, "L'intégration des secteurs de l'électricité au Canada," *L'Idée fédérale*, March 2012.

7. Jean-Benoît Nadeau, "Hydro-Québec : la bataille du Nord-Est," *L'actualité*, March 15, 2010.

8. David Maher, "The road to Gull Island" *The Telegram*, December 12, 2018.

Chapter Ten. Of Motors and Batteries

1. Karine Limoges, "L'expertise québécoise en motorisation électrique vendue aux Américains," *Électricité Plus + Le magazine*, July 12, 2018.

2. Bertrand Marotte, "Hydro-Québec on a research quest for the 'God Battery,'" *The Globe and Mail*, February 5, 2016.

Chapter Eleven. Hydro World

1. Maëlle Turbide, "Hydro-Québec s'entend avec le Réseau de transport d'électricité d'Europe," *Le Devoir*, November 29, 2018.

2 46. Ibid.

3. Denis Lessard, "Hydro fait du surplace," *La Presse+*, June 21, 2017.

Chapter Twelve. The Reactive Customer

1. *Annual Report 2017*, Hydro-Québec, March 2017, 6–7.

2. Hydro-Québec also signed seven contracts for small private generating stations producing a total of 107 megawatts, plus 12 contracts for cogeneration plants producing a total of 227 megawatts.

Chapter Thirteen. Visible from Space

1. David Abel, "In Québec, it's power versus people on hydroelectricity," *The Boston Globe*, January 23, 2018.
2. Jessica Nadeau, "Les pieds dans les tourbières de la rivière Romaine," *Le Devoir*, August 21, 2018.
3. Charles Lecavalier, "Des centaines de sites pollués par Hydro-Québec à nettoyer," *TVA Nouvelles*, November 19, 2017.

Chapter Fourteen. Not in My Back Yard

1. Julie Barlow, "La bataille de Saint-Adolphe-d'Howard," *L'actualité*, May 15, 2016.
2. Fleury, *Les coureurs*, 275–277.
3. Jean-Patrice Desjardins, "La MRC des Laurentides contre-attaque," *Journal Accès*, February 23, 2016.

Chapter Fifteen. Masters in Their Own House

1. Delphine Jung, "Le premier avocat cri veut briser les préjugés," *Droit-Inc*, March 12, 2018.
2. Jean-Benoît Nadeau, "Abolir la loi sur les Indiens ?," *L'actualité*, October 15, 2004.
3. The Association would become the Assembly of First Nations Quebec-Labrador in 1977.
4. Stéphane Savard, "Les communautés autochtones du Québec et le développement hydroélectrique: un rapport de force avec l'État, de 1944 à aujourd'hui," *Recherches amérindiennes du Québec*, 39, nos. 1–2 (2009): 47–60.
5. Benoît Aubin, "Le beau risque de Ted Moses," *L'actualité*, January 1, 2002.
6. Jean-Benoît Nadeau, "La Dernière frontière," *FP500/La Presse*, June 2002.
7. "La Société Makivik et la FCNQ signent une entente historique pour créer une nouvelle entreprise spécialisée dans le développement des énergies renouvelables au Nunavik," Société Makivik, February 21, 2018.
8. Savard, "Les communautés autochtones," 47–60.
9. Jean-Benoît Nadeau, "La vérité sur l'entente avec les Innus" and "L'entente en 12 questions," *L'actualité*, December 15, 2002.
10. Six of the nine communities sit formally on the Board of Directors for the partnership, but all nine communities have signified their intention to participate in it.
11. Denis Lessard, "Le président d'Hydro-Québec soulève l'ire du gouvernement," *La Presse*, August 9, 2018.
12. Hélène Baril, "Ni mirage, ni Klondike" and "Potentiel hydroélectrique du Québec," *La Presse*, November 27, 2018.

13. Hugo Joncas, "Les mystérieux barrages de Jean Charest," *Les affaires*, January 31, 2009.

Chapter Sixteen. Political Highs and Lows

1. Brian Myles, "Hydro-Québec: curieux bras de fer préélectoral," *Le Devoir*, August 10, 2018.
2. Terence Corcoran, "Boondoggle: How Ontario's pursuit of renewable energy broke the province's electricity system," *The Financial Post*, October 6, 2016.

Energy or Power?

Energy and power are two fundamental notions in electricity that are often confused. But they are very distinct. Energy is the quantity of electrons that flow over a given period of time—a second, an hour or a year. Power is the intensity of this flow at any given moment (usually a second).

Electrical power is measured in watts (or kilowatts, mega-, gigawatts). Energy is measured in watt-hours (or kilowatt-hours, mega-, giga-, tera-). There's no such thing as a Hydro-Québec generating station that supplies 1000 megawatts of *energy*: it supplies 1000 megawatts of *power*.

A generating station supplies 1000 megawatt-hours of energy if it functions at 1000 megawatts for an hour (or 100 megawatts for 10 hours).

At home a microwave oven requires 1000 watts to function. That's enough power to light ten 100-watt light bulbs (or a hundred 10-watt light bulbs). If many of those light bulbs are lit at the same time, they require a lot of power. But unless they are lit for a long time, they don't consume very much energy.

For example, one light bulb of 1 watt requires very little power. But if it is lit for 1000 hours, it will consume 1000-watt-hours (or one kilowatt-hour). And that costs 6¢ in Quebec. A 1000-watt stove burner will use 1 kilowatt-hour if it runs for an hour. Even though the power is different in the two examples, the consumption of energy is the same and both will still cost 6¢.

Hydro-Québec power stations that generate more than 245 megawatts

Name	Watercourse	Generating capacity
Robert-Bourassa	La Grande River	5 616
La Grande-4	La Grande River	2 779
La Grande-3	La Grande River	2 417
La Grande-2-A	La Grande River	2 106
Beauharnois	Lake Saint-François et Beauharnois Canal	1 900
Manic-5	Manicouagan River	1 596
La Grande-1	La Grande River	1 436
René-Lévesque (Manic-3)	Manicouagan River	1 326
Jean-Lesage (Manic-2)	Manicouagan River	1 229
Bersimis-1	Betsiamites River	1 178
Manic-5-PA	Manicouagan River	1 064
Outardes-3	Aux Outardes River	1 026
Sainte-Marguerite-3	Sainte-Marguerite River	882
Laforge-1	Laforge River	878
Bersimis-2	Betsiamites River	845
Outardes-4	Aux Outardes River	785
Eastmain-1-A	Eastmain River	768
Carillon	Outaouais River	753
Romaine-2	La Romaine River	640
Toulnustouc	Toulnustouc River	526
Outardes-2	Aux Outardes River	523

Name	Watercourse	Generating capacity
Eastmain-1	Eastmain River	480
Brisay	Caniapiscau River	469
Romaine-3	Romaine River	395
Péribonka	Péribonka River	385
Laforge-2	Laforge River	319
Trenche	Saint-Maurice River	302
La Tuque	Saint-Maurice River	294
Beaumont	Saint-Maurice River	270
Romaine-1	Romaine 2 River	70

Source: Hydro-Québec.

Small Glossary of Electric Terms

Advanced home automation
The use of electronic and digital resources and telecommunications to improve human habitats.

Alternator
Apparatus (also called a generator) that produces alternating current through electromagnetic induction by way of a stator and rotor.

Ampere
Unit of measurement of the flow of electric current in a conductor.

Brownout (load shedding)
Systematically cutting current or lowering voltage for specific times in periods where electrical energy is restricted.

Dam
A barrier constructed with valves, spillways and penstocks that are used to control water level from upstream, to direct water toward another river or to regularize its flow.

Energy
In electricity, measured in watt-hours. It corresponds to the amount of power (watts) used over the period of an hour. Energy occurs in electromagnetic, thermal, mechanical, chemical or radiation forms (See Appendix 1).

Head race canal
A canal that carries part of the flow of a river toward a generating station.

CANDU
Abbreviation of Canada Deuterium Uranium. A nuclear reactor that uses natural uranium as fuel (in the form of fuel pellets of uranium dioxide) and heavy water (deuterium oxide) as a moderator and coolant.

CO_2 equivalent
Value of reference used to express the quantities of emission of different greenhouse gases (GHGs). Uses carbon dioxide as a benchmark. Makes it possible to compare different types of energy according to their potential for global warming.

Cost price
Cost determined by calculating the fees necessary to build, produce and supply a good or service.

Diversion
The act of diverting the direction of a canal or river temporarily or definitively.

Dyke
A long wall or embankment built to prevent flooding from the sea. Contrary to a dam, a dyke has no mechanisms for controlling water flow.

Floodgate
Mobile waterproof partition used to let water out of a dam.

Generating station / Hydropower plant
A power plant that has the equipment to generate electricity by converting other types of energy (the mechanical force of water, in the case of hydroelectricity).

Ghost charge
Energy consumed by electronic and electrical appliances that are switched off but plugged into a power outlet (also called "phantom load," "standby power" and "vampire energy").

Gigawatt
A unit of power (GW) equal to one billion (1,0000,000,000 or 10^9) watts. The biggest generating station in Quebec, Robert-Bourassa, in James Bay, generates 5.6 gigawatts (or 5600 megawatts) of power.

Gigawatt-hour
A unit of energy (GWh) equivalent to a billion (1000,000,000) watt-hours. Corresponds to the annual energy consumption of about 50 single-family houses.

Kilovolt
Unit of voltage (kV) equivalent to one thousand volts.

Kilowatt-hour
Unit of energy (kWh) equivalent to 1000 watt-hours. The energy used by a 1000-watt wall heater over the course of an hour.

Megawatt
A unit of power (MW) equivalent to 1 million watts. Approximately the power required to light and heat one hundred houses.

Megawatt-hour
Unit of measure for energy (MWh) equivalent to 1 million watt-hours. An average single-family house consumes between 16 and 18 megawatts of energy per year.

Penstocks
Canals located between the head race canal and the generating station that carry water to a turbine. The speed of the water is significantly increased by the fall of the water before reaching the turbine.

Power
The total quantity of electricity supplied in a given moment (generally measured in seconds). Expressed in watts, power corresponds to the combined effect of voltage (measured in volts) and current (measures in amperes).

Reservoir
A large natural or artificial basin of water built to store, regularize or control the use of water.

Rotor
The mobile part of an alternator.

Spillway
Apparatus in dams or hydraulic works which is used to release excess water.

Stator
The immobile par of an alternator.

Substation
Installation at the junction of transmission lines, subtransmission lines and distribution lines that modifies voltage, current and other characteristics. It also protects the system from risk of damage.

Terawatt
A unit measurement of power (TW) equivalent to 1000 billion or one trillion (10^{12}) watts. A terawatt corresponds to more than 20 times the installed capacity of Hydro-Québec.

Terawatt-hour
Unit of measurement of power (TWh) equivalent to 1000 billion or 1 trillion (10^{12}) watt-hours. A city the size of Sherbrooke, Quebec, consumes 2 terawatt-hours of energy per year.

Transformer
An apparatus for reducing or increasing the voltage of an alternating current.

Turbine
The rotating part of a machine that receives the energy of a fluid (water, steam) and transforms it into mechanical energy. The turbine sets an alternator in motion, which induces an electrical current by the action of the rotor, in the stator.

Turbine Generator Unit
When the turbine rotates, it activates an alternator composed of a rotor (that turns) and a stator (immobile). The rotor is equipped with magnets and the stator with copper wires. The rotation of the rotor's magnets produces (or induces) the electrical current in the copper wires.

Volt
Unit of measurement of electric potential (V). The current of a house is 110 volts, but certain appliances, like ovens or driers, are at 220 volts. In the network, voltage varies from 600 to 735,000 volts.

Voltage
The difference in electrical potential between two points.

Watts

Unit of measurement of power (W). The prefixes kilo-, mega-, giga- and tera- indicate 1000, 1 million, 1 billion and 1 trillion watts. The small light bulb of a nightlight uses 4 watts.

Watt-hour

Unit of measurement of energy (Wh) equivalent to the energy consumed over an hour by the power of a watt. A light bulb of 100 watts lit for an hour uses 100 watt-hours.

Bibliography

(This list excludes journalistic articles quoted in the text, as well as official documents of Hydro-Québec.)

Christine Beaulieu, *J'aime Hydro*, Montréal, Atelier 10, 2017, 235 pages.

André Bolduc, Clarence Hogue, Daniel Larouche, Hydro-Québec: *L'Héritage d'un siècle d'électricité*, Montréal, Libre Expression, 1989, 341 pages.

Alain Chanlat, *Gestion et culture d'entreprise: le cheminement d'Hydro-Québec*, avec la collaboration d'André Bolduc et de Daniel Larouche, Montréal, Québec Amérique, 1984, 250 pages.

Philippe Faucher et Johanne Bergeron, *Hydro-Québec, la société de l'heure de pointe*, Montréal, Les Presses de l'Université de Montréal, 1986, 221 pages.

Jean-Louis Fleury, *Les coureurs de lignes: histoire du transport de l'électricité au Québec*, Montréal, Stanké, 1999, 551 pages.

Jean-Guy Genest, *Godbout*, Sillery, Septentrion, 1996, 390 pages.

Pierre Godin, *René Lévesque, Héros malgré lui (1960-1976)*, Montréal, Boréal, 1997, 736 pages.

Howard Hampton, *Public Power*, Toronto, Insomniac Press, 2003, 298 pages.

Hydro-Sherbrooke, *Rapport annuel 2017*, Ville de Sherbrooke.

International Energy Association, "Power markets and electrification," World Energy Outlook 2017, Paris, OECD Publishing, Paris/IEA.

Carol Jobin, *Les enjeux économiques de la nationalisation de l'électricité (1962-1963)*, Montréal, Éditions coopératives Albert Saint-Martin, 1978, 205 pages.

KPMG, Analyse économique des centres de données, Analyse présentée à la direction Développement des affaires—Québec, Hydro-Québec, July 11 2017.

Marcel Labelle, *Adélard Godbout, précurseur de la Révolution tranquille*, Montréal, Lidec, 60 pages.

Roger Lanoue and Normand Mousseau, *Maîtriser notre avenir énergétique. Pour le bénéfice économique, environnemental et social de tous*, gouvernement du Québec, 2014.

Lars Persson and Thomas P. Iangerås, *Transmission Network Investment across National Borders: The Liberalized Nordic Electricity Market*, Research Institute of Industrial Economics, October 15 2018, 37 pages.

Pierre-Olivier Pineau et Simon Langlois-Bertrand, Électricité – Structures et options tarifaires (the me 1)—Balisage des structures et options tarifaires des distributeurs d'électricité et pistes de solution, Rapport présenté à la Régie de l'énergie, December 15 2016.

Pierre-Olivier Pineau, *L'intégration des secteurs de l'électricité au Canada: Bonne pour l'environnement et logique sur le plan économique*, L'Idée fédérale/The Federal Idea, March 2012.

Richard Rhodes, *Energy, A Human History, New York*, Simon & Schuster, 2018, 465 pages.

Stéphane Savard, "Les communautés autochtones du Québec et le développement hydroélectrique : Un rapport de force avec l'État, de 1944 à aujourd'hui," *Recherches amérindiennes au Québec*, volume 39, numéros 1-2, 2009.

Stéphane Savard, *Hydro-Québec et l'État québécois*, 1944-2005, Sillery, Septentrion, 2013, 452 pages.

Sustainable Development Solutions Network et Institute for Sustainable Development and International Relations, Deep Decarbonization Pathways Project (2015), Pathways to deep decarbonisation 2015 report-executive summary, SDSN-IDDRI.

Vérificateur général du Québec, Rapport du Vérificateur général du Québec à l'Assemblée nationale pour l'année 2018-2019.

Johanne Whitmore and Pierre-Olivier Pineau, État de l'énergie au Québec 2018, Chaire de gestion du secteur de l'énergie, *HEC Montréal*, décembre 2017.

Articles by the authors

On Hydro-Québec
- Les sept travaux d'Hydro-Québec, *L'actualité*, March 6, 2019.
- Hydro-Québec à la conquête du monde, *L'actualité*, May 15, 2016.
- Les 6 leçons du verglas, *L'actualité*, January 1, 2018.
- La Bataille de Saint-Adolph d'Howard, *L'actualité*, May 15, 2016.
- Une histoire électrisante, *Québec Science*, November 2012.
- Hydro-Québec: La bataille du Nord-est, *L'actualité*, March 15, 2010.

- Le Retour des grands barrages, *L'actualité*, February, 2004.
- Hydro prend la route, *L'actualité*, May 1, 2003.
- La météo «renifleuse», *L'actualité*, May 1, 1998.
- Volts sous contrôle—Le Bunker Hydro, *Québec Science*, November 1997.
- Le plan Caillé, *L'actualité*, October 1, 1997.
- Hydro-Québec au bord de la dénationalisation, *L'actualité*, May 1, 1997.
- Le géant Hydro-Québécois et les dix nains, *Plan*, October 1990.
- Hydro-Québec: Vermont's power source to the north, *Vermont Business*, October 1990.

On Energy

- Pourquoi la conférence de Paris ne sera pas un échec, *L'actualité*, November 24 2015.
- L'or noir en 22 questions, *L'actualité*, April 1, 2013.
- Vous dites «hydrolienne» , *L'actualité*, September 9, 2010.
- Bon jusqu'à la dernière goutte, *Québec Science*, May 2008.
- L'auto branchée, *Québec Science*, June 2003.
- Les Terre-Neuviens se tâtent..., *L'actualité*, June 1, 2002.
- Touche pas à Hydro One!, *L'actualité*, June 1, 2002.
- Wô les moteurs à deux temps!, *Québec Science*, June 1998.
- Les bébés Hydros, *L'actualité*, June 15, 1991.

Sujets connexes

- Des emplois venus du froid, *L'actualité*, May 1, 2013.
- Au nord du 49e, tout est gros, gros, gros, *L'actualité*, September 15, 2012.
- Y a du génie québécois là-dedans!, *L'actualité*, December 1, 2006.
- La vérité sur l'Entente avec les Innus, *L'actualité*, December 15, 2002.
- L'entente en 12 questions, *L'actualité*, December 15, 2002.
- La dernière frontière, *FP500/La Presse*, June 2002.
- Faut-il perdre le Nord ?, *Affaires Plus*, June 1994.
- La faillite du dinosaure atomique, *L'actualité*, March 1, 1993.
- Prochaine sortie : BAIE-JAMES, *L'actualité*, September 1, 1992.
- Le gourou des microcentrales, *Plan*, October 1990.

List of people interviewed

This list includes individuals not quoted but excludes those who requested anonymity.

Serge Abergel
Tom Adams
Pierre Arcand
Robert Baril
Simon Bergevin
André Bolduc
Mathieu Boucher
Michel Clair
Jonathan Côté
André Dagenais
Alexandre Deslauriers
Vincent Desormeaux
Viviane de Tilly
Rebecca Diamond
Martin Dufour
Rubens Durocher
Vincent-Michel Duval
Hugues Fortin
Fabrice Fossaert
John Gaspo
Guillaume Gilbert
Marie-Josée Gosselin
Éric Hamel
Sophie Hamel-Dufour
Guillaume Hayet
Marie-Jacinthe Hemsas
Hugo Jeanson

Francis Labbé
Patrick Labbé
Richard Lagrange
France Lampron
Roger Lanoue
Christian Laprise
Guy Lavoie
François Legault
Éric Martel
Jocelyn Millette
Francis Monette
Normand Mousseau
Paul John Murdoch
Pierre-Olivier Pineau
Yves Poissant
Marc-Antoine Pouliot
Anne-Marie Prud'homme
Patrice Richard
Martine Rodrigue
Christian Royer
Alain Saladzius
Gary Sutherland
Jean-Pierre Tardif
Olivier Tremblay
David Vincent
Karim Zaghib

Credits

Photographs of Hydro-Québec

Fig. 1-A

Fig. 1-E

Fig. 2-A

Fig. 2-B

Fig. 2-E

Fig. 5-A

Fig. 6-A

Fig. 6-B (below)

Fig. 6-D

Fig. 8-C

Fig. 9-A

Fig. 9-B

Fig. 9-D

Fig. 10-A

Fig. 10-C

Fig. 11-B

Fig. 12-B

Fig. 12-A

Fig. 13-B

Fig. 14-C

Fig. 14-D

Fig. 15-A

Fig. 16-B

Index

ALSO FROM BARAKA BOOKS

Storming the Old Boys' Citadel, Two Pioneer Women Architects of 19th Century North America by Carla Blank and Tania Martin

The Complete Muhammad Ali by Ishmael Reed

A Distinct Alien Race, The Untold Story of Franco-Americans by David Vermette

The Einstein File, The FBI's Secret War on the World's Most Famous Scientist by Fred Jerome

Montreal, City of Secrets, Confederate Operations in Montreal During the American Civil War by Barry Sheehy

Patriots, Traitors and Empires, The Story of Korea's Struggle for Freedom by Stephen Gowans

Israel, A Beachhead in the Middle East, From European Colony to US Power Projection Platform by Stephen Gowans

A People's History of Quebec by Jacques Lacoursière and Robin Philpot

The Question of Separatism, Quebec and the Struggle Over Sovereignty by Jane Jacobs

The First Jews of North America, The Extraordinary Story of the Hart Family, 1760-1860 by Denis Vaugeois

FICTION

Exile Blues, A Novel by Douglas Gary Freeman

Fog, A Novel by Rana Bose

The Daughters' Story, A Novel by Murielle Cyr

Yasmeen Haddad Loves Joanasi Maqaittik by Carolyn Marie Souaid

Printed by Imprimerie Gauvin
Gatineau, Québec